Supply Chain Configuration

Charu Chandra • Jānis Grabis

Supply Chain Configuration

Concepts, Solutions, and Applications

Second Edition

 Springer

Charu Chandra
University of Michigan-Dearborn
Dearborn, MI, USA

Jānis Grabis
Institute of Information Technology
Riga Technical University
Riga, Latvia

ISBN 978-1-4939-3555-0   ISBN 978-1-4939-3557-4   (eBook)
DOI 10.1007/978-1-4939-3557-4

Library of Congress Control Number: 2016931594

Printed on acid-free paper

This Springer imprint is published by Springer Nature
The registered company is Springer Science+Business Media LLC New York

To my parents and family

Charu Chandra

To my parents, Jānis and Inta,
and my family

Jānis Grabis

Preface

A configurable (hence also reconfigurable) system by definition can be redesigned and remodeled for specific applications for the new (or changed) environment, and upgraded rather than replaced. With a reconfigurable system, new products and processes can be introduced with considerably less expense and ramp-up time. Reconfiguration efficiency attributed to such systems can be achieved only by means of intelligent decision-making (i.e., use of system synthesis, analysis, and simulation). The supply chain for this system must also be configured, aided, and supported by information systems that enable all supply chain members to learn about these changes expeditiously and adjust their processes accordingly.

Supply chain management deals with complex interactions among supply chain members and decision-making problems. Whether to establish a supply chain configuration or reconfigure an existing supply chain is one of the major decisions to be made. Configuration defines the operating basis of the supply chain. Other managerial decisions are made using the elaborated configuration as input. Therefore, configuration decisions are subjected to particularly comprehensive evaluation, which in turn, requires utilization of a variety of models and tools. This book covers these models and tools with particular emphasis on model integration and combination.

The supply chain configuration problem in this book is perceived as determining which units (e.g., suppliers, plants) to include in the supply chain, their size and location, and establishing links among the units. In the wider sense, the configuration problem may also include designing and modifying supply chain control structures, information systems, and organizational structures. Such a focused approach allows for thorough coverage of problems, issues, and solutions such as configuration under demand uncertainty, impact of the supply chain power structure, and hybrid modeling.

Explicit focus on the configuration problem, in-depth coverage of configuration models, emphasis on model integration, and application of information modeling techniques in decision-making are distinguishing characteristics of this book.

The primary objectives of this book are:

- To establish a focused scope definition of the supply chain configuration problem.
- To develop a supply chain configuration framework supporting development of configuration models for specific cases.
- To discuss models and tools available for solving configuration problems.
- To emphasize the value of model integration to obtain comprehensive and robust configuration decisions.
- To propose solutions for supply chain configuration in the presence of stochastic and dynamic factors.
- To illustrate application of techniques discussed in applied studies.

An illustrative supply chain configuration case study is introduced in Chap. 2 of the book and is further elaborated in the subsequent chapters. The case study is used to exemplify utilization of various decision-making techniques discussed in the book.

Book Organization

This book is divided into four parts, which are devoted to:

- Defining the supply chain configuration problem and identifying key issues.
- Describing solutions to various problems identified.
- Proposing technologies for enabling supply chain configuration.
- Discussing applied supply chain configuration problems.

The contents of the book are organized in a 16-chapter format as follows:

Part I. Supply Chain Configuration Problem and Issues

Chapter 1. Configuration

This chapter describes the general nature of configuration. It talks about configurable (reconfigurable) systems, their need, focus, motivation, properties (or characteristics), and general issues and problems faced by configurable systems. Basically, this chapter is intended as an introduction to the "nature of configuration" before delving into the more specific supply chain configuration systems.

Chapter 2. Scope of Supply Chain Configuration Problem

Supply chain configuration is one of the principal supply chain management decisions. It has profound impact on other subsequent managerial decisions. This chapter aims to position supply chain configuration decisions as part of the overall supply chain management decision-making process and to define the scope of the configuration problem. The positioning is described by analyzing the typical sequence of decisions made in the supply chain environment: definition of strategic objectives → product selection → establishing the supply chain → strategic supply chain management → tactical supply chain management → operational supply chain management. The scope definition describes objectives of supply chain configuration, questions being answered, and parameters and costs involved. Alignment of configuration objectives with strategic objectives of enterprises involved in a supply chain, and the supply chain as a whole, is also analyzed.

Chapter 3. Literature Review

The supply chain configuration has been widely studied by both academicians and practitioners. This chapter reviews these studies and identifies common characteristics of the supply chain configuration problem. The existing research is categorized according to data used in decision-making and several criteria characterizing the decision-making problem and its environment. These criteria include the modeling approach used, application area, problem size, and others. Results of the literature review are used in defining focus areas of remaining chapters in the book.

Chapter 4. Reconfigurable Supply Chains: An Integrated Framework

The purpose of this chapter is to describe "reconfigurable supply chains," their need, and their advantages. Then, we lay out an integrated framework for their implementation that maps problems and issues with suggested methods and techniques (either published in the literature or those laid out in later chapters). Basically, it lays the foundation for methodology in Chap. 5 and solutions described in Part II of the book.

Chapter 5. Methodology for Supply Chain Configuration

Supply chain configuration is a multiple-step process. This chapter identifies methodological steps involved in this process and provides guidelines for accomplishing these steps. The methodology relies on the integrated reconfiguration framework introduced in Chap. 4 and the methods used for performing various steps of the methodology are elaborated in Part II of the book.

Part II. Solutions

Chapter 6. Knowledge Management as the Basis of Crosscutting Problem-Solving Approaches

The importance of this chapter is to highlight that solutions to supply chain configuration problems must integrate complex modeling and analysis techniques drawn from a host of disciplines, such as systems science, management science, decision sciences, operations research, systems engineering, industrial engineering, and information systems. A proper knowledge of management support to decision-making is required to handle such a cross-sectional approach. Taxonomical and ontological approaches to knowledge management are described.

Chapter 7. Conceptual Modeling Approaches

Information modeling is used to gain understanding about a decision-making problem, to formalize the decision-making problem, and to prepare input data for quantitative modeling. Enterprise modeling techniques representing complex organization using an interrelated set of modeling views are used for conceptualization purposes. Process modeling is used to gain understanding of a decision-modeling problem by describing entities involved and their interactions. Data modeling is used to describe decision variables, parameters, and constraints.

Chapter 8. Mathematical Programming Approaches

Mathematical programming is the most prominent tool used in supply chain configuration, specifically for establishing the supply chain network, because of its ability to deal with spatial issues effectively. This chapter presents the generic mixed-integer programming model used in configuration. Application of this model, computational issues, and modifications of the generic model are also discussed. This chapter also briefly discusses nonlinear, dynamic, and stochastic programming formulations of the configuration problem.

Chapter 9. Simulation Modeling and Hybrid Approaches

Simulation models are used in evaluating supply chain configuration decisions because of their ability to represent the problem realistically and to capture a wide range of factors. They can also be applied to select the most appropriate configuration from a limited set of alternative configurations. This chapter describes the characteristic features of simulation models used in supply chain configuration. Issues of validation of simulation models in the context of supply chain configuration are raised. An approach for automated model building in the framework of integrated decision modeling is discussed.

The integrated application of mathematical and simulation models leads to hybrid modeling, combining optimization and simulation aims to inherit advantages and to avoid disadvantages. Application of hybrid modeling in supply chain configuration is described. Two important hybrid modeling approaches are described: (a) optimization and simulation models are used sequentially, where optimization is used to establish the configuration and simulation used for comprehensive evaluation of this configuration; and (b) simulation-based optimization procedures, where the optimization model receives input data from the simulation model at each iteration. An automated approach to building hybrid models on the basis of common data models is presented.

Part III. Technologies

Chapter 10. Information Technology Support for Configuration Problem Solving

Information Technology (IT) has a major impact on supply chain configuration. IT services are used to find the most appropriate supply chain configuration (decision support), as well as to ensure operations of the established configuration (infrastructural support). The decision support side is implemented on the basis of the supply chain configuration conceptual model. The integrated supply chain configuration framework developed in Chap. 4 is implemented using an integrated supply chain configuration decision support system.

Chapter 11. Data Integration Technologies

Supply chain configuration problem-solving relies on availability of accurate data. The modern web and data mining technologies allow for accumulation and processing of vast amounts of data. The chapter describes the application of data integration technologies by bringing together data from heterogeneous sources and structuring these data in a way suitable for supply chain decision-making. It also illustrates possibilities for data driven supply chain configuration, if there is limited upfront structural information about the supply chain.

Chapter 12. Mobile and Cloud Based Technologies

Information processing velocity in supply chains has increased dramatically thanks in part to mobile and cloud based technologies. This chapter demonstrates that many modern supply chains are a combination of physical and virtual supply chain units. It proposes methods for evaluating the combined supply chains and introduces a new concept of cloud chains.

Part IV. Applications

Chapter 13. Application in Hi-Tech Electronics Industries

Supply chain configuration decision-making techniques are applied in many different industries. This chapter is one of the three chapters discussing supply chain configuration applications. It follows the supply chain configuration methodology and investigates supply chain configuration challenges at a contract manufacturing company, manufacturing electronic circuits and boards. The supply reliability is the main challenge explored in this case study.

Chapter 14. Application in ICT Distribution

The supply chain configuration methodology is also applied to study a case of an ICT Wholesaler and Distribution company. The main challenge analyzed is finding an appropriate configuration when entering a new market. Configuration decisions involve selection of the appropriate delivery contracts.

Chapter 15. Application in Health Care

Health care supply chains have a lot of potential for their continuous improvement, especially in the light of cost pressures. This chapter analyzes opportunities for applying supply chain management best practices in the health care industry. An e-health care supply chain model is discussed and an example of the hospital laboratory supply chain is investigated.

Chapter 16. Future Research Directions in Supply Chain Configuration Problem

The concluding chapter, which lays out the agenda of future research directions for the field as seen by the authors, is presented.

Changes in the Second Edition

The second edition has been largely rewritten. Although the flavor of earlier edition has been retained, added emphasis has been placed on the most recent theoretical developments and empirical findings in the areas of supply chain management and related topics.

Chapter 9 "Simulation Modeling Approaches" and Chap. 10 "Hybrid Approaches" in the first edition have been combined into Chap. 9 "Simulation Modeling and Hybrid Approaches" in the second edition, where techniques for simultaneous and integrated application of simulation and optimization approaches have been described.

A new part (Part III: Technologies) has been introduced whose focus is to introduce readers to various technologies being utilized for supply chain configuration. Chapter 11 "Information Technology Support for Configuration Problem Solving" in the first edition has been moved as Chapter 10 of this part in the second edition. In addition, this part has two new chapters: Chap. 11 "Data Integration Technologies" and Chap. 12 "Mobile and Cloud Based Technologies."

Part III: Applications in the first edition has been renumbered as Part IV. In addition, this section has been entirely rewritten with applications in hi-tech industries, ICT distribution, and health care described in Chaps. 13, 14, and 15, respectively.

Finally, various illustrations and references have been updated to reflect the current state of the art in research, throughout the new edition.

Target Audience

The book is targeted to a broad range of professionals involved in supply chain management. It is modularly structured to appeal to audiences seeking a discussion of theoretical and qualitative supply chain configuration problems or a description of more technical quantitative and computational problems, as well as those interested in applied supply chain configuration problems.

The main target group is graduate students in industrial engineering, systems engineering, management science, decision analysis, logistics management, operations management and applied operations research, and practitioners and researchers working in fields of supply chain management and operations management who aim to combine mathematical aspects of problem-solving with the use of modern information technology solutions.

Professional/technical readers. This category includes research directors, research associates, and institutions involved in both the design and implementation of logistics systems in manufacturing and service-related projects. Examples include the National Center for Manufacturing Sciences and the Southwest Research Institute.

Managers, product and process engineers, logistics coordinators, and production planners within the product design, manufacturing, and logistics departments of various companies will also find the book a useful resource.

Academic readers. Professors and research associates within universities and colleges in industrial engineering, manufacturing engineering, mechanical engineering, automotive engineering and engineering management, management science, and production and operations management will find the book interesting to read.

This book may be used for teaching in graduate and professional development courses. It is also a valuable reference material for research in the area of supply chain management, logistics management, and operations management. The professional societies interested in these areas are:

- Institute of Industrial Engineers (IIE).
- Society of Manufacturing Engineers (SME).
- IEEE.
- INFORMS and Engineering Management Society.
- Production and Operation Management Society (POM).
- Decision Sciences Institute (DSI).
- American Production and Inventory Control Society (APICS).

Dearborn, MI   Charu Chandra
Riga, Latvia   Jānis Grabis

Acknowledgments

We gratefully acknowledge all those who helped us in bringing this book to publication. First and foremost, we have greatly benefited from the wealth of a vast array of published materials on the subject of supply chain and associated topics, including supply chain management and supply chain configuration. We would like to thank Baiba Rajecka for her help with technical editing. We are particularly grateful to our industrial partners Aleksandrs Orlovs, Ilmars Osmanis, Peteris Treimanis, Eriks Eglitis, Aleksandrs Belugins, and Armands Baranovskis for sharing their insights on supply chain configuration practice.

We would like to thank the reviewers of this book. The content of this book has benefited immensely from their valued insights, comments, and suggestions.

We offer our grateful appreciation to our families who have shown enormous patience and showered their encouragement and unconditional support throughout the long process of completing this book project.

Finally, we wish to thank the editorial staff in applied sciences area and the entire Springer production team for their assistance and guidance in successfully completing this book.

Contents

About the Authors

Charu Chandra is a professor in College of Business at the University of Michigan-Dearborn. Prior to this, he was a professor in Industrial and Manufacturing Systems Engineering at the same institution. He has worked in the industry as an information technology manager and systems analyst. He is involved in research in supply chain management, and enterprise integration issues in large complex systems. Specifically, his research focuses on studying complex systems with the aim of developing cooperative models to represent coordination and integration in an enterprise. He has published extensively in leading research publications in the areas of the supply chain management, enterprise modeling, information systems support, inventory management, group technology, and health services design and optimization. He has guest edited several journal issues on many of the aforementioned topics. He also serves on the editorial boards of several academic journals in many of the areas mentioned above. He teaches courses in decision sciences, business analytics, information technology, operations research, and supply chain management. He has a Ph.D. in industrial engineering and operations research from the Arizona State University. He is a member of Institute of Industrial Engineers, Institute of Operations Research and Management Sciences, Production and Operations Management Society, Association for the Advancement of Artificial Intelligence, and Academy Health.

Jānis Grabis obtained his Ph.D. in Information Technology from Riga Technical University in 2001. He spent 2 years with the University of Michigan-Dearborn and is currently a professor at Riga Technical University and leads the Department of Management Information Technology. His main research interests are supply chain management, enterprise applications, simulation, enterprise integration, and software project management. He is the author of more than a hundred scientific publications. He is a member of Decision Science Institute and IEEE.

Part I
Supply Chain Configuration Problem and Issues

Chapter 1
Configuration

1.1 What Is Configuration?

Modern organizations operate in a continuously changing environment influenced by economic, political, social, and technological developments. These dynamics of change have presented business enterprises with unprecedented opportunities and challenges in their quest for finding new ways to compete. Firms are beginning to move from operating on a regional or national to a global scale. They are increasingly replacing the traditional hierarchical organizational structure with centralized control to a flexible, decentralized setup with varying degrees of autonomy. They are striving to offer customized products in specialized markets to stay competitive. The ability to quickly adapt to changes, such as with time-to-market products, as well as incorporate institutional reforms will be the key to survival for firms.

In this environment, products are reaching a large consumer population across different market segments with expectations of high quality, low cost, and large product variety. This is resulting in increased complexity in all phases of the product life cycle, as well as rapid turnaround of products. With shorter life cycles than before, the need for product innovation has never been greater. With consumers so demanding, pre as well as post sale service have assumed extra significance, and even a source of competitive advantage. One of the primary means employed by firms to achieve innovation is *configuration,* defined as follows:

> "Configuration is an arrangement of *parts* or elements that gives the *whole* its inherent form."[1]

This definition points to the fact that configuration is achieved through a calibrated perturbation of *system* elements aimed at meeting a revised set of functional requirements and objective(s) for the product, as the core of its existence.

---

[1] Merriam-Webster Dictionary and Online Thesaurus.

© Springer Science+Business Media New York 2016
C. Chandra, J. Grabis, *Supply Chain Configuration,*
DOI 10.1007/978-1-4939-3557-4_1

The primary catalysts for achieving configuration are the product–process–resource interactions.

In this book, we present and explain concepts, solutions, and applications that are important for the effective configuration of the supply chain. The supply chain, which is also referred to as the logistics network, represents an integrated system. It consists of; (a) *entities,* such as suppliers, manufacturers, warehouses, distributors, and retailers, and (b) their *relationships* as they manage the flow of materials in the form of raw materials, work-in-process, and finished goods inventories. To optimize the performance of this system, it is essential to configure it based on the changing dynamics of supply and demand in the market. Before we look into various aspects related to configuration of the supply chain throughout the book, let us first define a configurable system, such as the supply chain and why it is needed, and some of the key issues in managing a configurable system.

1.2   What Is a Configurable System?

In defining configuration, the relationship between the whole–part components describes a system in its most basic representation. Because configuration changes the form of the whole, it can be described as a manifestation of a system at any given state relative to its original state at the time of conceptualization. Configuration affects a system's functions, either marginally or completely altering its form. Usually, the basis of configuration is the desire to upgrade or improve the functionality of the system. A system that embodies these dynamic properties is a *configurable* system. We propose a configurable system approach that integrates a system's components from concept to feasible solution. These are:

> *System and system design concepts → system of systems → sources of configuration (product–process–resource) → sources of configuration (public policies) → configuration problems → configuration models → configuration solutions*

We describe these elements next.

1.2.1   System and System Design Concepts

A configurable system, as a class of system, follows a general system's main traits but has its own unique features. It is based upon the following three main principles:

- Principle 1—A configurable system is based upon a *whole–part* relationship
- Principle 2—A configurable system encapsulates interdisciplinary knowledge
- Principle 3—The General Systems Theory (GST) influences the design of a configurable system

1.2.1.1 Whole–Part Relationship

A general system stands for a set of things (or entities) and the relationship among these things. Formally, we had $S = (T, R)$, where S,T,R denoted a system S, a set of things T distinguished within a domain S, and relation (or relations) R defined on T, respectively. Thing (T) consists of seven components: $T = (I, O, E, A, F, M, P)$, which are input (I), output (O), environment (E), agent (A), function (F), mechanism (M), and process (P), respectively. These components of a generic system are described below in Table 1.1 (Nadler 1970).

Formally, the system (whole) may be defined as an assemblage of subsystems (parts), and agents and mechanisms (people, technology, and resources) designed to perform a set of tasks to satisfy specified functional requirements and constraints. In a configurable system, *parts* may define its physical, logical, and virtual systems. For example, these may represent the manufacturing, logistics, and Internet (or eCommerce) systems (or subsystems), respectively. For a configurable system the *whole* gives it form, structure, organization, and arrangement.

Relationships are defined among system components and can be both internal among system elements (identified in Table 1.1) and external with the system's environment. The level of control exerted on the system (i.e., at the strategic, tactical, and operational levels) also defines relationships.

Systems give organization a formal structure, a purpose, a goal (objective), and above all a basis for integration. Such a structure is beneficial for an organization in managing its complexity, integration of its functions, and aligning its product–process–resource structure. System also provides the framework that an organization needs for designing and implementing models, methodologies, tools, and techniques for aligning its business (es) and improving productivity.

In the light of the above explanations, it can be construed that a configurable system is a specialist system, which combines to yield a *system-of-systems* that performs the function of an integrated system for the entire product life cycle, i.e., from concept generation to its maturity.

Table 1.1 System components

System component	Examples
Input	Physical item, information, or service that is necessary to start processes
Output	Physical item, information, or service that results from processing of input. The output is related to the total accomplishment of the function
Environment	Physical or sociological factors within which system elements operate. It relates to resource requirements, both physical and human
Agent	Computational or human resources for carrying process
Function	Mission, aim, purpose, or primary concern of the system
Mechanism	Physical or logical facilitators in the generation of an output
Process	Flows, transformations, conversions, or order of steps that transforms an input into an output

1.2.1.2 Interdisciplinary Knowledge

Ludwig von Bertalanffy formulated a new discipline, General System Theory (GST) (Von Bertalanffy 1968, 1975), and defined its subject matter as "formulation and derivation of those principles which are valid for systems in general whatever the nature of the component elements and the relations or forces between them." GST enunciated the principle of unification of science, and its essence was interdisciplinarity. It produced a new type of scientific knowledge—interdisciplinary knowledge. According to Bertalanffy, there is some element of isomorphism (state of similarity) that allows extension of one scientific discipline to other sciences. Thus, in complex systems such as the configurable system, we see the design of knowledge at a high level or generic level, and low level or the domain level. These are, therefore, labeled as general knowledge and domain (specific or expert) knowledge.

1.2.1.3 Influence of GST on System Design

The biggest influence that GST has had on system design is in its formalization. For example, a system is designed to recognize its whole–part relationship instantiated in its environment (both internal and external).

The concept of isomorphism has facilitated system design by recognizing similarity (or commonness) across entities, relationships, and environmental variables. Similarity implicitly recognizes relationships, thereby improving a system's representation and eventually impacting its performance (quality, reliability, etc.).

Another useful feature of GST in system design is separating information needs (and associated knowledge) at the domain independent (or generic) level from that of domain dependent (or specific/problem) level. Such an approach ensures that the system captures both breadth and depth of knowledge. Because the latter is embedded in the former, the captured knowledge has a larger context, thereby ensuring interactions and thus larger relevance. It also ensures that the knowledge does not become redundant. In Table 1.2, we provide a brief explanation of various design principles that play a part in the overall design of a configurable system.

1.2.2 Sources of Configuration

As the definition of configuration given in the earlier section suggested, it affects a system's characteristics, such as form, structure, organization, and arrangement. The system's product, process, and resource dimensions mainly represent the sources of these characteristics. We discuss these next.

- *Product-related configuration* is usually implemented as a result of implementation of strategies that make:
  - Changes in product characteristics, such as adding more variety due to changes in newer models, colors, additional user-friendly features, etc.

Table 1.2 Key design principles for configurable system design

Design principle	Explanation
Unity	All systems (and their components) are whole (unity) depending on the context in which they are represented
Commonality	All systems in the universe of systems share common universal characteristics
Isomorphism	Similarity (and therefore commonality) among system components and associated relationships
Reuse	Commonality leads to reuse and eventually standardization, conformity, and reliability
Abstraction	Enables managing complexity by abstracting features of system's components. It also allows representation of relationships, such as whole–part and generalization–specialization
Polymorphism	Creates classes of systems and reuses them for specialized functions
Encapsulation	Enables encapsulating knowledge and information-hiding on objects (and classes) to create uniqueness of objects (and classes)
Independence	Domain independent vs. domain dependent knowledge creation
Inheritance	Enables the avoiding of information redundancy and information-hiding by clustering information representation where they rightfully belong

- Changes in product specifications as a result of either new or enhanced functional requirements due to customer needs, performance standards, process changes, and service criteria
- Changes in product structure as a result of changes in product design for manufacture, assembly, delivery, new processes, and technology employed for product development

- *Process-related configuration* is implemented as a result of improved or enhanced process technology that enables the enterprise to achieve agility and flexibility in their manufacturing operations, as well as integrate various processes. It enables the achievement of modularity in product development and the acquisition of specialization.

- *Resource-related configuration* is implemented in response to the requirement of specialized, knowledge-intensive resources by the enterprise as it adopts newer advanced technologies to improve its performance.

- *Organization-related configuration* is implemented to meet the need for enhancing organization controls as the decision-making process is carried out in an enterprise. Such a situation arises as decentralized, semiautonomous, or autonomous decision-making is introduced to improve the quality and speed of decision-making.

- *Service-related configuration* is implemented with a view to improving and maintaining both prior and post product delivery service in a customer-centric environment.

- *Competitive strategy-related configuration* is implemented as a result of strategy adoption, such as off-shoring, outsourcing, mass customization, time-to-market, and globalization that have the potential of offering a competitive advantage to an enterprise.

- Others

  - *Change in lead-time.* Product development can potentially be highly integrated and, as such, any change in lead-time for any product component will involve reconfiguring the system to account for its impact.
  - *Change in pricing.* This may impact sales contracts and revenue sharing contracts among the enterprise partners due to a potential change in product sales volume or its market share.
  - *Change in location* on either production or delivery for any component of the product life cycle will be cause for reevaluation, and hence configuration of the system. It will particularly affect production and logistics activities because these involve movement of goods and associated transportation activities.
  - *Change in supplier selection* either to add or remove a supplier must be accounted for in the product development process. Such a decision may have a major impact on product quality, product development, production scheduling, etc.
  - *Change in product or process cost* may occur due to changes in the cost of procuring raw materials and other technologies required in the delivery of products.
  - *Change in contracts.* Revenue sharing, cost sharing, technology sharing, and resource-sharing arrangements are entered into between enterprise and its business partners.

1.2.3   Impact of Public Policies on Configuration

Many of the social, economic, political, environmental, and technological developments of our times are driving configuration in systems. Public policies enunciated by governmental and nongovernmental organizations and industry, which monitor or regulate industrial and business practices, are one of the primary means of implementing suitable changes or reforms. Among some of the significant policy issues with major impact on business practices, and consequently on configuration of systems, are as follows:

- *Energy conservation.* Consumption of natural fuel in automobiles, for industrial production, household appliances, and utilities
- *Health care reforms* and their impact on total business costs
- *Social security* entitlements for seniors and their impact on national economy
- *Water and natural resource management,* especially due to increasing consumption by the rising global population
- *Biotechnology* and its impact on problems in business, engineering, and medical sciences
- *Nanotechnology* for unique applications to major problems in engineering, science, and medicine

- *Cybersecurity* for data integrity and security among various supply chain entities and in their interactions in decision-making
- *E-commerce* for the manner in which various supply chain entities have to adopt to doing its business
- *Cloud Computing* and its impact on how data and information could be shared with unique and diverse Internet implementation strategies
- *Social networking* and its impact on consumer preferences and market sentiments about the products profiled in the supply chain

Public policies are capable of having major impacts at national, regional, and local levels and, as such, solutions designed for these problems must recognize both global and local implications. Accordingly, the factors considered for evaluation in models designed for problem solving are chosen carefully and deliberately.

1.2.4 Configuration Problems

There are two types of problems that are encountered in a configurable system.

At the *macro* level, the whole system is considered and decisions concerning all entities involved are made. Generally, strategic decision making problems are addressed at this level. Examples of such problems include the following:

- How much to invest in new or existing plants or warehouses and at what locations?
- Which products to include or exclude in the existing product portfolio?
- Which services to implement or enhance to improve the product experience?

At the *micro* level, issues concerning implementation of the macro level decisions are addressed. Examples of such problems are

- How much plant capacity should be allocated to a particular product?
- Which product(s) are to be scheduled for production on a given machine?
- Which products should be stored or warehoused at a certain warehouse location?

The problem of coordination and synchronization of activities and resource utilization occurs at all levels of implementation in an enterprise. A common problem encountered is that of *information sharing* among various members/partners in an enterprise. This often leads to either under or overutilization of resources and impacts scarce resources, such as capacity and inventory.

1.2.5 Configuration Models

Similar to configuration problems, configuration models may also be classified into two types.

Macro model. It describes behavior of the whole system with emphasis on strategic decision-making. Models used are characterized by higher level of abstraction and generality.

Micro model. At this level, the models are designed to investigate behavior of individual entities involved in the system. These models are domain dependent and are designed to solve specific problems.

A third type of model, a *coordination model,* is usually designed to coordinate the interactions between macro and micro level models. This is typically by way of arriving at solutions that meet the objectives of the two types of models.

1.2.6   Configuration Solutions

Configuration solutions designed for solving configuration problems are closely aligned to the configuration models. Examples of some of these solutions are:

- *Configuration network optimization* that is aimed at maximizing the revenue flow throughout the network
- *Global optimization* that attempts to optimize both functional and interorganizational objectives
- *System integration* through collaborative planning among various enterprise partners
- *Customer value* for service level maximization through statistical planning and control, and total quality management techniques
- *Information technology and decision support systems.* Implementing enterprise resource planning decision support systems for collaborative planning throughout the enterprise.

These and other solution techniques are discussed throughout this book.

1.3   Why Is a Configurable System Needed?

The motivation for developing configurable systems is the desire to use advanced systems for complex problem solving that can be designed, modeled, and configured according to specifications suitable for specific applications—flexibly and with agility, and upgraded and reconfigured rather than replaced. With a reconfigurable system, new products and processes can supposedly be introduced with considerably less expense and ramp-up time.

The emphasis of configuration is purely on *focused changes* to the system, rather than its total redesign. The changes are caused by many of the sources of configuration and policies described in the previous section. The notion of focused changes is based on incremental or additive design, which implies that a design may not be done from scratch. Instead, an existing design case may be used as the basis with the proviso of refinement/revision for the final designed product.

1.4 Examples and Applications of Configuration

The concept of configuration has been widely applied across many fields and in several different applications. In each case, however, as the definition of configuration given in Sect. 1.1 suggests, it symbolizes the notion of arrangement of parts (components or subsystems) forming a whole (system). Implicit in this definition is the idea of integration towards a common purpose or objective. Below, we give examples from various disciplines where the concept of configuration has been applied successfully to improve/upgrade systems.

In *computer systems,* a configuration is deemed as an arrangement of functional units according to their characteristics. Often, configuration refers to the choice of hardware or software, or combination of both. For instance, a configuration for a personal computer consists of main memory, a hard disk, communication devices, external memory drives, a LCD monitor, and the operating system, among other components. Many software products require that the computer have a certain minimum configuration. For example, the software may require a graphics display monitor and a video adapter, a specific microprocessor, and a minimum amount of main memory. Similarly, when a new device or program is installed, it sometimes needs to be configured, which means setting values of parameters. For example, the device or program may need to know what type of video adapter is available and what type of printer is connected to the computer. Similarly, computers are configured to interface with Internet, mobile communication features and other advanced software for business and social computing needs. The next level of configuration is connecting to various service providers such as online search and document management service providers and procuring computer support services. Many of these services can be replaced as necessary. On the other hand, external service providers may change their service provision conditions which results in a need for reconfiguring the system.

In the *building construction industry,* configuration refers to the structure and form of the building, such as a dome or an apartment building.

In many industrial applications, configuration refers to the change in physical layout. For example,

- *Airfield runway* layout and configuration refers to the maximum possible number of aircraft landing and takeoff due to the layout of the runway
- *Refinery plant and facilities.* Each petroleum refinery is uniquely configured to process specific raw material(s) into a desired line of products
- In *mathematics,* the concept of configuration space is utilized in defining the position of a single point in an n-dimensional plane
- In *mechanical engineering,* it is possible to tailor an engine configuration for a certain operation or operations
- In *chemistry,* molecules can be configured according to certain structural arrangements and properties
- In *atomic physics and quantum chemistry,* the electron configuration is the arrangement of electrons in an atom, molecules, or other bodies

- In *logistics systems,* which span a supply chain, configuration refers to the choice of locations for either production or warehousing, or both, and how to organize raw materials and other goods inventory to support various echelons in the supply chain. Obviously, this is the primary topic of this book and a detailed discussion on supply chain configuration follows in the remainder of the chapters
- In *health care systems*, operating rooms are configured to meet the specific needs of various surgery procedures
- In *networking systems*, cloud network offers unique ways for computing on virtual servers
- Many social systems, provide the ability to configure communication and interaction among like-minded

1.5   Key Issues in Configuration

There are key issues encountered in developing configurable systems, and their impact is felt across all levels (i.e., strategic, tactical, and operational) of decision-making in an enterprise.

1.5.1   Coordination and Synchronization

In a configurable system, there is a high level of integration among its parts (or components). This integration is achieved through common strategies and policies, and objectives for the whole (system). In order to achieve it, a high degree of coordination and synchronization of plans and actions among the parts is required.

1.5.2   Conflicting Objectives

Various parts (components) that together define the whole (system) have their own objectives. As we configure them together, invariably these objectives come in conflict or work against each other. For example, the objective of minimizing costs in one subsystem may be at odds with maximizing product variety in another subsystem. It becomes quite important, therefore, to find a compromise between these conflicting objectives.

1.5.3   Complex Network

The structure and functioning of a configurable system may become highly complex, especially when the subsystems (plants or facilities) are co-located and there is a high-level of interlinking among them (e.g., flow of materials or inventories occurs within the plant or facility). Obviously, the question arises on the makeup of the structure so it will meet the stated objectives (of both parts and whole), which in this case may be shortest lead-time or least cost.

1.5.4   System Variation Over Time

A configurable system is a self-adapting, dynamic system. As described earlier, this could happen due to changes in any of the system's components or controls exercised via various strategies or policies reflecting changes in the environment. For example, if the demand input to a production system is based on the point-of-sale data captured through various order entry outlets, the configured system would naturally integrate inputs and outputs from all related subsystems (i.e., forecasting, order management, inventory management, production planning, and shipping and warehousing).

1.5.5   Push–Pull Strategies

One of the ways business enterprises have remained competitive is by pushing change through turnover of products and their inventories so that when the consumer demands shifts, the system is nimble and agile enough to respond to changing circumstances. By adopting an approach to work on a push–pull strategy, they are able to postpone adoption of emerging changes to products and associated processes and/or resources in the product life cycle, as late as possible without adversely affecting the business. This is achieved by pushing the product in the product life cycle until such phase or time, that it could be easily pulled away, in order to reflect evolving changes resulting in a new product configuration.

1.5.6   Direct-to-Consumer

With the advent of the Internet and its creative uses in all aspects of the product life cycle, it is quite natural for enterprise systems to be configured to shorten the time required for a product to reach the ultimate consumer. This implies the elimination of echelon(s) throughout the product life cycle. Some noteworthy examples are application of eCommerce techniques, the role of traditional middleman (such as a travel agent in the airlines industry, teller in the banking sector, order taker in consumer and mail-order catalog industries) is either being eliminated or becoming irrelevant and, therefore, unnecessary. The end result is that manufacturers or suppliers are reaching the end-consumer directly, thereby realizing savings in time and cost.

1.5.7   Strategic Alliance

As products are being designed to offer enhanced features, it is becoming apparent to firms that they do not have the capability to go it alone. They are, therefore, seeking strategic alliances by partnering with other firms who add value to the product, and help meet the targeted objectives. However, the resulting arrangement raises more questions, primarily related to synchronizing the plans, strategies, and objectives of alliance partners, as well as sharing the common benefits among them.

1.5.8   Mass Customization

One of the ways firms have attempted to differentiate their products to consumers is by offering customized products. This has been achieved by designing products that meet conflicting objectives of low cost, high quality and customer value, large variety, and shorter lead times. The challenge lies in how to, (a) configure various systems that support product life cycle to absorb variations in consumer demands, and (b) the resulting activities to support their fulfillment without causing a major disruption in the enterprise system.

1.5.9   Outsourcing and Procurement Strategies

As firms find innovative ways to compete, they have resorted to strategies that would bring down product costs and/or lead times. Outsourcing of components, business functions, and services are increasingly being used as means to achieving these strategies. Such activities, however, lead to a major problem in coordination of ordering and receiving so that the product may be assembled or produced according to schedule.

1.5.10   Information Technology and Decision Support Systems

These have played an important role as enablers of various functions, as well as decision-making tools in the product life cycle. This is particularly true in the case of enterprise resource planning systems that firms have been using successfully to integrate actions and policies across functions and entities in an enterprise. The problem, however, arises whenever newer functions are introduced into the enterprise, especially when various parts or entities of the system are either not ready or incapable of integration due to various reasons, primarily lack of technological capabilities.

1.5.11 Customer Value

This should be measured in tangible or intangible terms. Intangible value can be measured by customer perception of the product in terms of usefulness, appeal, etc. Tangible value can be measured by price, after sales service, warranty, etc.

In the rest of the book, we describe various solution approaches and techniques to many of the above issues. These utilize models and algorithms drawn from operations research, statistics, simulation, and information sciences disciplines among others.

References

Nadler G (1970) Work design: a system concept. Irwin, Homewood
Von Bertalanffy L (1968) General system theory. George Braziller, New York
Von Bertalanffy L (1975) Perspectives on general system theory. George Braziller, New York

Chapter 2
Scope of Supply Chain Configuration Problem

2.1 Introduction

As firms position themselves to stay competitive, they face the challenge of transforming their operations from a static to a dynamic business environment. An obvious choice for transformation is supply chain operations because of their potential impact on almost every aspect of the business encompassing the extended enterprise. This is a complex undertaking because supply chain management entails managing the following under the umbrella of a common framework:

- Entity relationships, such as product, process, resource, organization, supplier, retailer, and customer
- Flow of goods, services, cash, and information
- Objectives, strategies, and policies

Further, the framework is developed to account for risk and uncertainty caused by factors internal and external to the enterprise. Obviously, this requires reconfiguring the supply chain in order to keep pace with the changing environment.

In this chapter, we focus on studying the nature of the supply chain and its configuration in a dynamic business environment. We develop an understanding of the basis for a supply chain configuration problem, its classifications, and its various dimensions. This chapter also introduces a running example of bicycle suppl0y chain used throughout the book to illustrate various concepts and methods.

© Springer Science+Business Media New York 2016
C. Chandra, J. Grabis, *Supply Chain Configuration*,
DOI 10.1007/978-1-4939-3557-4_2

2.2 Supply Chain and Supply Chain Management

Companies deliver products and services in response to the customer demand. In order to produce and deliver products, companies procure services and materials from their partners. As a result, a partnership network of the companies is established. The main characteristics of the network are (1) the flow of products starting with materials used in production to the ready to use end-products, (2) the flow of information about customer demand and coordination of production and delivery activities, and (3) dependence of all companies involved in the network on satisfaction of the end-customers. The definition below captures all three afore-mentioned facets of supply chains:

> Supply chain is a network of supply chain units collaborating in transforming raw materials into finished products to serve common end-customers.

A supply chain unit is defined as an entity involved in the supply chain and having a distinct legal or spatial identity. Although each unit can perform multiple roles in the supply chain, they usually have a type characterizing the main purpose of the unit such as manufacturing plant, distribution center or warehouse. All supply chain units can belong to one company as in the case of vertically integrated supply chains or supply chain units belonging to different companies.

The supply chain units are linked together in multiple ways forming a network structure. Nevertheless a chain-type of superstructure is imposed by movement of materials and products from their initial state to the final state in the form of end-products. The links are mainly used to represent physical movement of products although the pervasive use of information technology makes this distinction vague because of developments like 3D printing (see Chap. 12). Multiple links are possible between any two units in the supply chain.

Figure 2.1 shows a graphical representation of SCC Bike supply chain, which is used as an example throughout the book (the SCC Bike case is described in Appendix). The supply chain is represented as a directed graph. The supply chain units and links are shown as the graph's nodes and edges, respectively. The unit type is indicated by a marker. Given that supply chains often have a large number of units, some of the units are represented in an aggregated manner using clusters. In this example, customers are divided into clusters according to their geographical location.

Systematic and predictable supply chain operations are achieved through rigorous supply chain management. Supply chain management is a coherent set of techniques for planning and execution of all supply chain management processes, enacting a chosen supply chain strategy and ensuring customer demand satisfaction. The supply chain strategy is derived on the basis of competitive strategies of companies involved in the supply chain, and there are four main supply chain

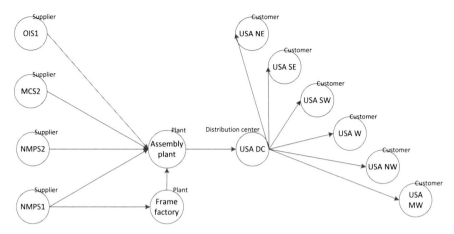

Fig. 2.1 Sample supply chain network

strategies, namely, lean, flexible, agile, and service-oriented. There is a large number of supply chain management processes described in Sect. 2.2.2. The key property of the supply chain management processes is their cross-enterprise and cross-sectional nature.

2.2.1 Supply Chain Management Strategies

The three typical supply chain management strategies are lean, flexible and agile strategies. The service-oriented strategy is added here to emphasize emerging importance of services including electronic services in supply chains. Companies often employ different combinations of strategies and hybrid strategies depending upon product and market segmentation, as well as other factors.

2.2.1.1 Lean

According to Vonderembse et al. (2006), "a lean supply chain employs continuous improvement efforts that focus on eliminating waste or non-value steps along the chain. It is supported by the reduction of setup times to allow for the economic production of small quantities; thereby achieving cost reduction, flexibility and internal responsiveness. It does not have the ability to mass customize and be adaptable easily to future market requirements." This type of supply chain is essentially based on the lean principles, which advocate the reengineering of business processes to remove all non-value added activity, generally ascribed as the source of waste in the system. Another significant feature of the lean technique applied in the lean supply chain is integration across functions of the enterprise.

The accrued benefits are a high capacity utilization rate, shorter lead times, and minimization of total supply chain costs. Jasti and Kodali (2015) have summarized these features as the pillars of lean supply chain management. The eight pillars are information technology management, supplier management, elimination of waste, JIT production, customer relationships management, logistics management, top management commitment, and continuous improvement.

From the supply chain network perspective, it is expected that the influence of lean supply chains focuses on establishing long-term links among the supply chain units and minimization of number of units and links. The lean supply chains aim towards simplifying and streamlining the supply chain network, while providing a high level of standardization and specialization.

2.2.1.2 Flexible

The flexible supply chain strategy addresses uncertainties associated with supply chain operations and primarily demand uncertainty. The flexibility is an ability in a relatively inexpensive way to respond to changes in customer demand and shift production and delivery to products with the highest demand and value. This ability usually is already built-in in the system, therefore, supply chain already should be designed to provide a certain level of flexibility. This characteristic limits a kind of changes and level of uncertainty the supply chain is able to react, and designing flexible systems usually is more expensive than designing lean systems.

From the supply chain network perspective, flexible supply chains have built-in redundancies and cushions in the form of extra units and links to deal with changes and uncertainties. The supply chain units and links are less specialized and multiple functions can be performed. It is argued that the flexible supply chain strategy attempts to deal with uncertainty without drastic overhaul of the supply chain network.

2.2.1.3 Agile

The agile supply chain strategy supplements the flexible supply chain strategy. However, it does not shy away from substantial changes in the supply chain network and attempts to introduce changes in a proactive manner. The agile supply chain configuration strategy is often described as a combination of flexibility and adaptability by reconfiguring the supply chain network. The key enablers of the agile strategy are collaboration among the supply chain units, advanced information technology and other technical capabilities and knowledge management (Gunasekaran et al. 2008). The key limitation of the agile strategy is the difficulty to balance agility and restrictions set by long-term investments in the supply chain network.

From the supply chain network perspective, agile supply chains are characterized by large variety of units involved, often performing fine-granularity functions.

2.2.1.4   Service-Oriented

The key feature of service orientation is provisioning of required capabilities and resources on-demand from service providers. The services are composed together to create a supply chain suited for current or expected business opportunities. The service-oriented approach minimizes fixed investments and ramp-up time. It relies on using advanced information technologies and cloud computing as discussed in Part III of this book.

From the supply chain network perspective, service-oriented supply chains have much less strong associations with particular spatial location of supply chain units and customers. More importantly, the primary focus switches from the physical movement of products to the electronic movement of information and delivery of services. The distinction between the physical and electronic worlds blurs in service-oriented supply chains.

2.2.2   Supply Chain Management Processes

The supply chain network describes the static structure of the supply chain while processes provide a dynamic representation of supply chain management activities. Supply chain management processes are cross-enterprise, cross-sectional, and self-similar.

The cross-enterprise processes involve multiple companies in the execution of supply chain processes. The important feature of these collaborative processes is that the companies involved are mainly concerned with their inter-communications rather than with internal operations of each supply chain unit. That simplifies development and execution of complex supply chain processes. The cross-sectional processes involve multiple supply chain problem areas such as sales, purchasing, and logistics. This characteristic implies that supply chain decision-making and process execution cannot be done in isolation and mutual interactions and dependencies among different problem areas should be taken into account.

The Supply Chain Operations Reference (SCOR) model (Supply Chain Council 2011) categorizes supply chain management processes in five groups: (1) plan, (2) source, (3) make, (4) deliver, and (5) return. The plan processes represent planning of supply chain operations. The source processes describe receiving of the products from preceding supply chain units. The make processes describe transformation of products at the supply chain unit. The deliver processes represent delivery of the products to consecutive supply chain units. The return processes represent reverse logistics activities. It suggests that these characteristic processes define the base dynamics of supply chain operations and they can be observed for different levels of aggregation of supply chain units. Therefore, the processes are referred to as self-similar. The processes can be further detailed at different levels of abstraction, making it possible to analyze cross-sectional supply chain characteristics.

2.3 Supply Chain as a System

A supply chain can be perceived as a social-technical system. The system is defined as a tuple:

System = <Components, Interrelationships, Boundary, Purpose, Environment, Input, Output, Interface, Constraints>

Refining the system's properties specifically to the supply chain case yields:

Supply Chain = <Supply Chain Units, Links, Boundary, Purpose, Environment, Input, Output, Interface, Constraints>

Components become Supply Chain Units as supply chains consist of supply chain units, and similarly Interrelationships are replaced by Links. The system boundary can be formally described as all links attached to the supply chain units without specifying their source or target. Logically that means connections with other supply chains and units outside of the scope of the given supply chain. The overall supply chain purpose is to serve its customers. The more detailed breakdown of supply chain objectives is given in Chap. 7. The supply chain environment is defined by its competitive environment. The supply chain inputs are materials and services provided by supply chain units outside the scope of the given supply chain, and outputs are products and services delivered to customers. Customers are often shown as final nodes in the supply chain networks and thus within the supply chain system. However, this representation of customers concerns only their physical location while their logical behavior is external to the supply chain system. The main supply chain Interface is customer demand. Supply chain constrains are classified as network wide constraints and unit wide constraints. The network wide constraints mainly define global operating requirements such as regional differences, legal requirements and others. The unit wide constraints define local operational requirements, such as allocation of resources, capacities, labor, and capital.

As for any other system, the key properties of the supply chain system are decomposition, modularity, coupling and cohesion. The supply chain can be decomposed starting with the top level network. As stated above, the top level network consists of units having distinct legal or spatial characteristics. The units are further decomposed to represent their internal structure. For instance, a warehouse consists of multiple docking places and subdivisions. The decomposition is related to different levels of supply chain decision making. There are decisions: (1) associated with the entire supply chain; (2) made at the unit level; and (3) made at the unit subdivision level. Similarly, supply chain decisions are categorized as strategic, tactical, and operational. The strategic decisions are made in the planning horizon measured in years; the tactical decisions are made in the planning horizon measured in months; and the planning horizon of the operational decisions is measured in weeks or shorter time units. Different planning horizons can be used at every decision-making level, e.g., there could be operational decisions made at the network level.

2.4 Supply Chain Management Problem Domain

Supply chain management involves dealing with multiple managerial and technical problems (Cooper et al. 1997; Mentzer et al. 2001; Soni and Kodali 2013). These problems highlight several common issues that must be addressed for a supply chain to function effectively and efficiently. We discuss below some of these issues and how they have been addressed in the published literature.

2.4.1 Key Challenges

Customer engagement. This issue takes a holistic view of sales and customer relationships. Customer relationships management and business analytics techniques are used to provide customized customer services on a global scale. From the supply chain perspective, it changes cost optimization focus to customer service focus implying not only high fill rates and responsiveness but also social responsibility, supply chain transparency, and environmental consciousness (Carter and Rogers 2008; Danese and Romano 2013).

Distribution Network Configuration. This issue deals with the selection of warehouse locations and capacities, determining the production level for each product at each plant, and finalizing transportation flows between plants and warehouses so as to maximize production, transportation, and inventory costs. This issue relates to information sharing: (a) inter-firm between marketing, production planning, inventory planning, and receiving and warehousing functions, and (b) intra-firm between manufacturer, suppliers, distributors/retailers, and transporters. It is a complex optimization problem dealing with network flows and capacity utilizations (Ballou 2001; Mangiaracina et al. 2015).

Inventory Management and responsiveness. This issue deals with stocking levels at various echelons in the supply chain. Demands from echelon-to-echelon are considered in making this decision. This is a decision problem solution which involves using algorithms for forecasting, and inventory management, in conjunction with simulation and optimization capabilities. Retailers, suppliers, and manufacturers deal with this issue in a supply chain by sharing information on customer demand, inventory levels, and replenishment schedules (Childerhouse et al. 2002; Sheffi 1985).

Supply Contracts. This issue deals with setting up relationships between suppliers and buyers in the supply chain through establishment of supply contracts that specify mutually agreed-to prices, discounts, rebates, delivery lead times, quality standards, and return policies. This approach differs from traditional approaches because its central focus is on minimizing the impact of decisions made at not just one echelon in the supply chain, but on all its players. A retailer sets up these contracts with a distributor or directly with a manufacturer. To manage this issue, it is incumbent upon various supply chain players to share information related to product price, cost, profit margins, warranty, and so on. This is a decision problem solution that could range from a simple linear programming problem to a complex game theory algorithm (Cachon 2002; Fisher et al. 1997; De Matta and Miller 2015).

Distribution Strategies. This issue deals with decisions pertaining to the movement of goods in the supply chain. Among the strategies available are direct shipments, cross-docking involving trans-shipments, and load consolidation. The objective is to minimize warehousing (storage) and transportation costs. A manufacturer makes decisions about either warehousing or direct shipment to the point of usage of various products, utilizing information shared among manufacturers, suppliers, distributors, and retailers in the supply chain. Solutions to this problem involve network algorithm utilizing linear, and nonlinear programming techniques in deterministic and stochastic environments (Frohlich and Westbrook 2001; Cagliano et al. 2008).

Supply Chain Integration and Strategic Partnering. One of the key issues in managing supply chains is integration (Bramham and McCarthy 2004). Information sharing and joint (or collaborative) operational planning are basic ingredients for solving this issue. Implementation of Collaborative Planning, Forecasting and Replenishment (CPFR) (Aviv 2001; Ng and Vechapikul 2002; Caridi et al. 2005; Fliedner 2003), as carried out by Wal-Mart retail stores in their supply chain aided by information sharing through common software platforms such as Enterprise Resource Planning (ERP) are viable strategies (Akkermans ct al. 2003). In a manufacturing supply chain, it would mean CPFR among the retailer, supplier, and the manufacturer of products. The main idea of this technique is to avoid carrying excess inventory through accurate forecasting, and utilizing commonly agreed to demand data, information about which is shared among various supply chain partners (Anonymous 2000).

Outsourcing and Procurement Strategies. An important issue to consider is what to manufacture internally and what to buy from external sources. One of the problems to be dealt with when making these decisions is identifying risks associated with these decisions and minimizing them. Another issue to consider is the impact of the Internet on procurement strategies and what channels to utilize (public or private portals) when dealing with trading partners. In arriving at the decision of whether to outsource or buy, various optimization models may be utilized to balance risk and payoffs. Once this decision has been made, use of appropriate information technology components, such as Internet portals and procurement software, plays a key role in these decisions. An example of this issue in a manufacturing supply chain may be the decision to outsource a component assembly rather than making it in-house. Information sharing for outsourcing and other procurement issues is accomplished in the supply chain and its extended enterprise, for intra-firm and inter-firm, via Intranet, Extranet, and Internet portals (Chen et al. 2004).

Information Technology and Decision Support Systems. One of the major issues in supply chain management is the lack of information for decision-making. Information technology plays a vital role in enabling decision-making via information sharing throughout the supply chain. Some of the key ingredients of information technology in the supply chain are use of Internet and Web-based service portals, integrated information/knowledge within ERP software, and decision support systems that utilize proven algorithms for various strategic, tactical, and planning problems in specific industry domains (Fiala 2005). Significant progress has been achieved in enabling physical supply chain integration. Lau and Lee (2000) use the distributed objects approach to elaborate on an infrastructure of integrated component-based supply chain information systems. Kobayashi et al. (2003) conceptually discuss workflow-based integration of planning and transaction processing applications, which allows for effective integrated deployment of heterogeneous systems. Verwijmeren (2004) develops the architecture of component-based supply chain information systems. The author identifies key components and their role throughout the supply network. Themistocleous et al. (2004) describe the application of enterprise application integration technologies to achieve physical integration of supply chain information systems. However, approaches and technologies for logical integration at the decision-modeling level, where common understanding of managerial problems is required, are developed insufficiently (Delen and Benjamin 2003).

Challenges for Information Sharing in the Supply Chain. In light of various decision-making levels and issues facing effective management of the supply chain, it becomes imperative to find globally optimal integrated solutions. However, it is difficult to achieve depending on whether the problem-solving models designed for the purpose achieve local (or sequential) or global optimization of the supply chain network. Depending on which approach is adopted, the requirement for information sharing will be starkly different. For example, in the case of sequential supply chain optimization, the objective of its individual partners is optimized without regard to the overall supply chain network objective. Accordingly, the need for information sharing is limited and/or closed, sometimes nonexistent and usually offline. For global supply chain optimization, however, the objective for the overall supply chain takes precedence over each partner's objective. For this scenario, information sharing is extensive, open, and online (Beamon 1998; Fiala 2005; Simchi-Levi et al. 2007).

2.4.2   General Problems

The main general supply chain management problems are:

Competitiveness. The house of supply chain management (Stadtler 2008) considers solving this problem as the ultimate goal of supply chain management. To maintain competitiveness, a supply chain must outperform competing supply chains in at least some aspects such as prices, quality, or delivery responsiveness.

Customer service. It characterizes the ability of supply chains to meet customer requirements. Approaches to addressing this problem are as diverse as the customer requirements representing such aspects as cost, quality, and responsiveness.

Coordination. Coordination of decisions by each supply chain member are made with regard to the impact these decisions will have on the performance of other supply chain members.

Collaboration. Joint activities performed by supply chain members to achieve common goals (Kliger et al. 2015) include product design and planning. In the case of collaborative product design; manufacturers, suppliers, and potential customers work together to design product that best suits market requirements and the capabilities of parties involved.

Environmental protection. Supply chains as a system operate and interact with its environment including they impact on nature and consumption of natural resources. Increasingly supply chain management decisions are made with regard to these concerns.

Flexibility and agility. Customer requirements and operating environments are dynamically changing. Addressing flexibility and agility issues implies the ability of reactive and proactive response to change.

Globalization. This presents both opportunities and challenges. Cost reduction and expansion in new markets have become possible. On the other hand, increasing competition, local regulations, and cultural adjustments cause additional difficulties.

Integration. Addressing the integration problem enables customer service improvements, coordination, and collaboration. Information sharing is an important integration sub-problem.

Mass customization and postponement. Customers demand individualized products with similar cost and delivery time characteristics as those of standardized products. Postponement is one of the strategies for delivering market-specific and customized products. It implies location (in time and space) of the product finishing close to the point of demand.

Outsourcing. Firms focus on their core competencies to achieve a high level of competitiveness in specific areas while allocating supporting functions to partners.

Risk/benefit sharing. Implemented supply chain decisions have different impacts on supply chain members. Some of the units may assume larger risks and incur additional costs in the name of overall supply chain benefit. Risk and benefit sharing is essential for building trust and enforcing commitment among supply chain members.

Robustness. Supply chains operate in uncertain environments. Operations need to be planned and executed with respect to this uncertainty.

Sustainability and social responsibility. Supply chains are designed and operated with regard to social, cultural, and environmental issues.

2.4.3 Specific Problems

The main specific supply chain management problems are:

Demand planning and forecasting. Demand data are required for other supply chain management activities. Demand planning attempts to influence demand to make supply chain operations more efficient.

Finance. In the supply chain management framework, this concerns planning of supply chain costs and controlling supply chain performance.

Inventory management and logistics. Problems deal with delivering products and services to customers, including planning of distribution structure, inventory management, warehousing, and transportation activities. Reverse logistics is employed to process customer returns and residue of other supply chain operations

Production planning and manufacturing. These problems address creation of products and services in response to customer demand. It includes such supply chain management concerns as master production planning, capacity allocation, scheduling, maintenance of manufacturing facilities, and manufacturing quality.

Marketing and sales. The primary concerns of these managerial problems are attracting customers and processing their orders.

Network design. A network of supply chain units meeting product and process design requirements is established. Problems to be addressed concern location and role of supply chain units, allocation of products, strategic-level capacity planning, and establishing transportation and information exchange links.

Process design. This is a significant supply chain management problem because of the very large number of processes that can be potentially enumerated as the supply chain is functionally decomposed top-down from a tier $\rightarrow$ unit $\rightarrow$ function $\rightarrow$ process level, and then need to be properly managed. One of the key problems that arise is how to develop a composite process design of the supply chain that clusters these processes based on similarities in features and characteristics, and arranges clusters according to an optimal implementation schedule.

Product design and bill of materials. This is not an explicit supply chain management problem, although there are significant interactions between design and logistics activities and at this stage, it is a major input for further supply chain management activities. From the supply chain management perspective, this problem concerns collaborative product design, balancing product design requirements and supply chain capabilities, and providing the bill of materials for further planning purposes.

Personnel management. Workforce requirements are considered while dealing with the personnel management problem. This includes workforce planning, hiring, layoffs, promotion, training, and incentives.

Supplier selection and purchasing. This deals with procurement of materials and services that are needed from suppliers to satisfy customer demand. The problem includes such issues as identification of materials and services needed, supplier relationships (i.e., supplier selection, contract negotiation, supplier evaluation) and execution of procurement operations.

2.5 Supply Chain Configuration

The supply chain design problem is one of the key supply chain management problems. It defines the underlying network structure of the supply chain and is interrelated with a number of other supply chain management problems such as logistics, purchasing, and others. The concurrent decision-making concerning supply chain network structure and key attributes of the supply chain at the network level is referred to as supply chain configuration. For the purposes of this book, the following definition is adopted:

Supply chain configuration is a set of supply chain units and links among these units defining the underlying supply chain structure and the key attributes of the supply chain network.

The supply chain configuration problem is to select appropriate supply chain units and to establish links among these units, as well as to make key decision concerning supply chain attributes at the network level. From the configuration point of view, every supply chain unit has specific geographical location and functions, and the links are mainly used for physical movement of goods from one unit to another. The configuration decisions are made according to the overall supply chain strategy. The appropriate supply chain units are selected from a number of alternatives depending upon the unit type.

A configurable supply chain is a system that efficiently adapts to its environment, offered in the form of supply and demand issues for the product(s) to be manufactured. A configurable supply chain is needed to manage logistics in a configurable system. This is because the adopted policies for product, process, and resource components of a configurable system have to be integrated with both inbound and outbound logistics decisions to realize benefits of flexible strategies. Some of the key triggers for designing and implementing a configurable supply chain are as follows:

- Introduction of new product(s), or upgrade for existing product(s)
- Introduction of new, or improvement in existing, process(es)
- Allocation of new, or reallocation of existing, resource(s)
- Selection of new supplier(s), or deselection of existing ones
- Changes in demand patterns for product(s) manufactured
- Changes in lead times for product and/or process life cycles
- Changes in commitments within or between supply chain members

A configurable supply chain can help in assessing the impacts of one or more of the following factors/activities in a configurable system:

- Flows due to materials, inventory, information, and cash
- Throughput due to movement of products
- Capacity utilization
- Costs at various stages of the product development life cycle
- Lead time in product development
- Batch and lot sizing
- Process redesign
- Product development strategies
- Procurement and/or allocation of resources
- Strategic, tactical, and operational policies for the supply chain

Analysis of these factors/activities involves dealing with a wide range of managerial problems and spans across all tiers of the supply chain. Problem-solving approaches need to consider both interactions among factors and activities, and supply chain members.

2.6 Supply Chain Configuration Dimensions

The supply chain configuration problem solving allows to define scope of the given supply chain system. The scope is defined along multiple dimensions, namely, horizontal extent, vertical extent, objectives and criteria, decisions made, and parameters.

2.6.1 Horizontal Extent

The supply chain is usually divided into tiers (or stages, or echelons). Each tier consists of units with the same general functionality. The concept of tier should be treated with care, however, as differentiation between tiers is often fuzzy and units can belong to multiple tiers. That has become even more profound as supply chains assume networked structures. Still, tiers help structure the supply chain configuration problem and facilitate identification of common features of supply chain units.

The typical supply chain tiers, which can be further decomposed, are

- Customer tier—the most downstream tier
- Distribution tier
- Manufacturing tier
- Supply tier—the most upstream tier

Demand for supply chain products or services originate at the customer tier and it is transmitted upstream along the supply chain (Fig. 2.2). In many cases, customer nodes in this tier are an aggregation of individual customers clustering in a particular geographical location.

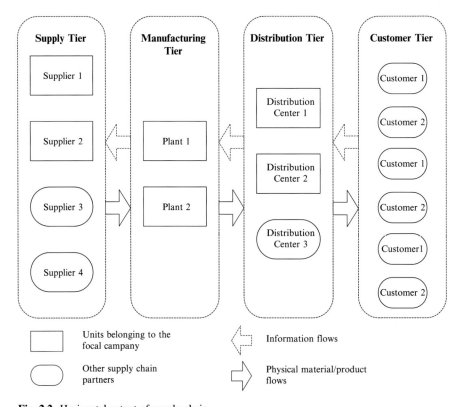

Fig. 2.2 Horizontal extent of supply chain

The distribution tier receives customer requirements and is responsible for delivering required products or services. It involves such general units as warehouses, distribution centers, and cross-docking points. These units are grouped into distribution sub-tiers. Alternatively, supply chain units in the distribution tier can be classified as wholesalers, retailers, and brokers. Third-party logistics providers present a special case for belonging to the distribution tier. In some situations, these can be represented by a single supply chain node.

There are two distinct scenarios to organize the supply chain's operations. The first, where manufacturing tier directly creates products or services demanded by the supply chain's customers. It receives demand information from the distribution tier. In return, it provides products to the distribution tier and orders materials from the supply tier. In the second scenario, the manufacturing tier can also be divided into several sub-tiers, such as preprocessing, assembly, final assembly, and finishing. Manufacturing outsourcing can be represented either in the manufacturing tier or in the supply tier. The first scenario is more relevant to representing the manufacturing tier for an engineering company such as Ericsson, which has outsourced almost all manufacturing operations and retained only product and process design as their primary competency, or in the case of capacity sharing agreements. The second scenario is more relevant for representation of manufacturing of components (for instance, the Ford and Visteon case).

The supply tier provides materials to manufacturing according to orders received. This tier can be divided into sub-tiers, linking raw materials suppliers, secondary suppliers, and direct suppliers. Representation of the supply tier depends upon the importance of supplied materials. Suppliers providing widely available and substitutable materials do not need to be represented by individual nodes.

A return tier could be treated as a separate tier in supply chains. It is responsible for handling customer returns and disposal of the returned products and waste. However, recycling of returns could occur at any of the identified core tiers and can be perceived as one of the integral processes performed along with the supply, manufacturing, and distribution activities.

One additional supply chain tier not sufficiently exposed in the literature is the utility tier. This tier includes providers of basic infrastructural services such as electricity, water, and recycling. That is of particular concern for global supply chains, because availability, cost, and quality of such services vary substantially.

Definition of this supply chain configuration dimension includes specifying the number of tiers in the supply chain, defining general types of units in each tier, and identifying specific constraints for the tier as a whole (for instance, the number of suppliers required).

2.6.2   Vertical Extent

As noted earlier, a supply chain consists of several members spread across many tiers. Each of the tiers consists of one or many business units (entities). Each of these business units is, by itself, an enterprise comprising functional areas such as design, marketing and sales, production planning and control, inbound and outbound logistics (procurement, receiving, warehousing, shipping), and so on. Each unit may also pursue its own independent strategies to manage its functions and strive to achieve specific goals and objectives.

A *within* unit (local) vertical integration would entail synchronizing and coordinating strategies and policies, for example, between its sales and marketing and manufacturing functions to achieve a common objective for the unit.

A *between* (global or supply chain level) vertical integration within a tier (comprising all units) would be to implement common strategies and policies to achieve a common (global) objective across units in their tier.

Vertical integration could be achieved at strategic, tactical, and operational levels of decision making within a tier of the supply chain. This is primarily achieved by means of implementing strategies and policies appropriate at these levels that are aimed at achieving long-term, mid-term, and short-term goals and objectives.

Definition of this supply chain configuration dimension includes specifying the number of units in each tier in the supply chain and identifying specific constraints and objectives: (a) within a unit at high level and by functional areas at low level, and (b) between units at high level and across functional areas at low level.

2.6.3   Objectives

Decision-making objectives are chosen according to general strategic objectives. Certain quantitative criteria or metrics are associated with each identified objective. General managerial concerns related to the supply chain configuration problem are

- What is the current supply chain performance?
- "What if" analysis?
- How to improve customer service?
- How to improve supply chain robustness and delivery reliability?
- Could supply chain be made more profitable?
- Is supply chain sufficiently flexible?
- How to improve cooperation?
- How to comply with local requirements?
- Whether to pursue outsourcing?
- Which partners to choose?
- Where to locate supply chain facilities?

Answering these questions leads to formulation of general supply chain configuration decision-making objectives. These objectives can be formulated on the basis of performance attributes identified in the SCOR model (Stewart 1997) as described in Table 2.1.

Table 2.1 Decision-making objectives

ID	Objective	Description
O1	To improve supply chain delivery reliability	The performance of the supply chain delivering the correct product, to the correct place, at the correct time, in the correct condition and packaging, in the correct quantity, with the correct documentation, to the correct customer
O2	To increase supply chain responsiveness	The velocity at which a supply chain provides products to the customer
O3	To increase supply chain flexibility	The agility of a supply chain in responding to marketplace changes to gain or maintain competitive advantage
O4	To optimize supply chain costs	The costs associated with operating the supply chain
O5	To improve supply chain asset management efficiency	The effectiveness of an organization in managing assets to support demand satisfaction. This includes the management of all assets—fixed and working capital

The aforementioned objectives usually can be expressed in quantitative terms and are used for decision-making on the basis of analytical models

2.6.4 Decisions

Initially, general supply chain configuration decisions are identified following the supply chain configuration decision-making objectives. These are subsequently specified using particular decision variables. Five groups of decisions are defined, characterizing structure, links, quantity, time, and policies used.

Structural decisions are

- Location of supply chain facilities at different tiers
- Facility opening
- Supplier selection
- Product allocation
- Definition of facility's capabilities

Decisions characterizing links among supply chain units are:

- Establishing a fixed link among a pair of units–if a link between units cannot be established on the spot, decisions must involve which units' link should be established
- Restricting cooperation to specified links–implies that a particular unit can cooperate only with a limited group of other units (i.e., a customer zone is served by only one particular distribution center)
- Choice of products or services delivery mode
- Choice of information exchange mechanisms

Alternative production location according to ownership, international/global, and product state are described by Meixell and Gargeya (2005).

Decisions characterizing quantity are:

- Quantity of purchased materials
- Quantity of products produced
- Quantity of products processed
- Quantity of products delivered
- Quantity of products stored in inventory
- Shipment quantities along supply chain links
- Capacity-related decisions

Decisions characterizing quantity often differ by their interpretation and level of detail. For instance, manufacturing capacity is specified for each product separately at a plant or for the entire plant. The main decision characterizing time is delivery time.

Decisions characterizing policies are

- Choices of manufacturing strategies. The most general values of these decisions are make-to-plan (make-to-stock), make-to-order, and assemble-to-order. The choice of the manufacturing strategy influences propagation of demand information along the supply chain and functions performed by different units
- Adoptions of information sharing policies. Information sharing policies affect manufacturing, inventory, and transportation, as well as several other decisions and characteristics. They also influence requirements towards information exchange infrastructure and adoption of common information exchange standards. Other IT-related decisions, such as implementation of ERP and manufacturing execution systems can also be considered
- Choice of distribution channels. Values these decisions assume include Internet-based distribution, third-party logistics, direct sales, quick response, continuous replenishment, and vendor-managed inventory. Some of the policies may be represented in relation to the horizontal extent dimension. For instance, the direct shipment policy implies the absence of intermediate distribution tiers. Multiple distribution strategies can be used in a single supply chain
- Choices of procurement policies. Some alternatives include volume consolidation, alliances and partnerships with suppliers, just-in-time (JIT), and manufacturing resource planning (MRP). From a technical perspective, various types of e-procurement can be chosen (for instance, EDI, Internet-based business-to-business (B2B) approaches, and trading networks)
- Adoption of outsourcing. Decisions apply to separate supply chain functions and indicate whether these are outsourced or not. That influences the way supply chain costs are accounted for. For instance, outsourcing may reduce fixed costs associated with a facility opening

Each of these policies can be parameterized by a set of particular structural, linkage, quantitative, and time parameters. For instance, if the decision is between using EDI or the Internet for information exchange purposes, a parameter characterizing a fixed cost for establishing links among manufacturing facilities and suppliers is larger for the first. Policies influence which supply chain management problems need to be addressed during decision-making. For instance, evaluation of the built-to-stock manufacturing strategy requires consideration of the inventory management problem.

The decisions listed above do not provide an exhaustive list of all supply chain configuration decisions. That, especially, applies to policy decisions. Decisions relevant to a particular decision-making problem, and decision variables characterizing these decisions, are defined during the supply chain configuration problem-solving process.

2.6.5 Parameters

Parameters usually are more specific to a particular decision-making problem compared to other supply chain dimensions discussed earlier. Some common features, however, can be identified.

Parameters are traditionally classified as internal and external. External variables for the supply chain configuration problem are customer demand and requirements in general, taxes, governmental regulations, and others.

The first group of internal variables represents structural characteristics, which includes representation of the existing supply chain structure, bill of materials, available capacity, and capacity requirements. This group also includes parameters describing attributes of alternative transportation channels (e.g., distance, speed).

Supply chain operations are described by cost- and time-related parameters. These are classified as fixed and variable parameters. Fixed cost parameters describe costs due to opening (closing) and operating supply chain facilities, capacity buildup costs, and costs associated with establishing and maintaining links among supply chain units. Inventory replenishment, manufacturing setup, and fixed transportation costs can also be considered. Variable costs are incurred per each processed product. Processing can assume various forms including transportation, assembly, inventory handling, and others. Parameters for representing processing time can also be used. Specific parameters may be needed to describe various attributes of the supply chain management policies considered.

2.7 Aligning Objectives

One of the major tasks of any supply chain configuration effort is to align the objectives according to several alignment perspectives (Table 2.2). The system perspective discussed in Sect. 2.3 concerns trade-offs among supply chain unit at various levels of aggregation, e.g., whole supply chain, partnership of individual units, or individual units. This issue is addressed using joint decision-making facilities and by considering profit and risk sharing among supply chain members.

Table 2.2 Methods for achieving alignment of supply chain configuration objectives

Alignment perspective	Alignment methods
System perspective	Joint decision-making
	Profit and risk sharing
Planning horizon	Multiple modeling views
Problem domain	Model integration
	Concurrent engineering

For instance, a metal processing company pays suppliers inventory carrying costs to prevent an inventory glut at the manufacturing site.[1] The planning horizon perspective concerns exploration of supply chain configuration decisions at strategic, tactical, and operational levels. Different factors are taken into account at each of the decision-making levels. Multiple modeling views at different levels of granularity are used to ensure that strategic decisions can be implemented at the operational level and that operational process are designed to support strategic decisions.

The problem domain perspective implies that supply chain configuration is shaped by the interplay of various general and specific supply chain management problems what is addressed by an increasing tendency to perform supply chain configuration concurrently with other managerial decisions. This increases modeling complexity although model integration helps to alleviate these issues. Multiobjective modeling is an approach for addressing all three alignment challenges.

2.8   Summary

In this chapter, we explore supply chain as a systems concept, and its configuration in the face of a dynamic business environment. We discuss various aspects of supply chain configuration problems, its classifications, and its various dimensions. We posit supply chain configuration as a supply chain management problem and argue that it can be successfully achieved if properly modeled around the decision-making levels and aligned with objectives formulated along different alignment perspectives.

The scope definition also includes identification of synergies and contradictions among configuration objectives along different alignment perspectives. It is suggested that key methods for achieving alignment are joint decision-making using integrated multi-view models, profit and risk sharing, multi-objective decision-making, and concurrent engineering.

---

[1] https://www.pnc.com/content/dam/pnc-com/pdf/smallbusiness/IndustrySolutions/Whitepapers/Driving_SupplyChain_Svgs_1110.pdf

Appendix. SCCB Case Description

SCC Bike is a sample bicycle manufacturing company and it is used throughout the book to illustrate various concepts and methods. The sample company reminisces real-life bicycle companies and the example is particularly influenced by the GBI and Shimano cases (Magal and Word 2012; Chang 2006). Factual information is derived from industry reports provided by Bicycle Retailer[2] and other professional publications.

The company is headquartered in Midwest USA, where it has a frame plant and an assembly plant. It manufactures medium to high end bicycles and offers around 20 different end-products manufactured out of around 250 parts. The manufacturing volume is around 250,000 bicycles a year. The bicycle production life-cycle is 150 days. Bicycle parts are categorized as non-moving, mechanical, and other industry parts. Examples of non-moving parts are saddle, rims, and handlebar. Examples of mechanical part are drivetrain, brakes, and gears. The other industries suppliers supply, for example, tires. A detailed bill-of-materials used in bicycle manufacturing is described by (Galvin and Morkel 2001). Parts are sourced from suppliers around the world and distributed through a set of specialist wholesalers and retailers. It is assumed that suppliers specialize in providing one specific category of parts and they provide a set of parts (as opposed to providing individual items). Suppliers are not necessarily manufacturers of the parts they provide. Specialist wholesalers and retailers collectively referred to as customers represent a specific sales area, the unit is responsible for.

The company's mission is to provide high quality bicycles at lower than premium pricing for cycling enthusiasts. It also focuses on providing professional customer care and direct collaboration with a limited number of specialist sellers. The demand for bicycles is growing continuously and supply chain configuration activities are driven by the need to provide an adequate customer service in new locations, which are currently only served by limited amount of bicycles sold over the Internet. Therefore, the main configuration objective is location of new distribution facilities and allocation of supplies to these distribution facilities. At the same time the manufacturing facilities are relatively stable. The company also continuously works with its suppliers to improve products and services. Some of the components can be sourced from several alternative suppliers.

Figure 2.3 shows the supply chain network of SCC Bike. The supply chain units in the figure are shown using their short names and Table 2.3 defines abbreviations used. The focal point of the supply chain is frame factory located in US MW and it supplies both assembly plants with hand-made carbon and aluminum frames. The suppliers deliver parts to the assembly plants, and each assembly plant is served by its regional set of suppliers. The customers are served from the regional distribution centers. The US DC is responsible for handling web sales. The SCC Bike has a bundling deal with cycling mobile app developers.

---

[2] http://www.bicycleretailer.com/

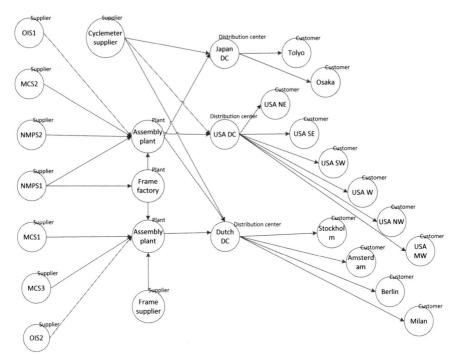

Fig. 2.3 The SCC Bike supply chain network

Table 2.3 Description of supply chain units for the SCC bike case

#	Short name	Name	Type	Location (state, country)
1	FF	Frame factory	Plant	WI
2	AP1	Assembly plant (US)	Plant	WI
3	CS	Cyclemeter supplier	Supplier	CA
4	MCS1	Mechanical components supplier	Supplier	ITA
5	OIS1	Other industries supplier	Supplier	NY
6	NMPS1	Non-moving parts supplier	Supplier	IL
7	MCS2	Mechanical components supplier	Supplier	CHI
8	MCS3	Mechanical components supplier	Supplier	TWA
9	NMPS2	Non-moving parts supplier	Supplier	CHI
10	US DC	US DC	Distribution center	WI
11	EU DC	Dutch DC	Distribution center	NL
12	JP DC	Japan DC	Distribution center	JPN
13	US NE	US NE	Customer	NY
14	US SE	US SE	Customer	FL
15	US SW	US SW	Customer	TX
16	US W	US W	Customer	CA
17	US NW	US NW	Customer	WA

(continued)

Table 2.3 (continued)

#	Short name	Name	Type	Location (state, country)
18	Tokyo	Tokyo	Customer	JPN
19	Osaka	Osaka	Customer	JPN
20	AMS	Amsterdam	Customer	NL
21	Berlin	Berlin	Customer	GER
22	Milan	Milan	Customer	ITA
23	STH	Stockholm	Customer	SWE
24	AP2	Assembly plant (EU)	Plant	NL
25	OIS2	Other industries supplier	Supplier	NL
26	NMPS3	Non-moving parts supplier	Supplier	NL
27	US MW	US MW	Customer	IL
28	GL WEB	Global web customers	Customer	

All supply chain units and links are also characterized by their attributes. These are introduced in the following chapters

References

Akkermans HA, Bogerd P, Yucesan E, van Wassenhove LN (2003) The impact of ERP on supply chain management: exploratory findings from a European Delphi study. Eur J Oper Res 146:284–301

Anonymous (2000) E-commerce will improve logistics. Hosp Mater Manag 25:2

Aviv Y (2001) The effect of collaborative forecasting on supply chain performance. Manag Sci 47:1326–1343

Ballou RH (2001) Unresolved issues in supply chain network design. Inf Syst Front 3:417–426

Beamon BM (1998) Supply chain design and analysis: models and methods. Int J Prod Econ 55:281–294

Bramham J, MacCarthy B (2004) The demand driven chain. Manuf Eng 83:30–33

Carter CR, Rogers DS (2008) A framework of sustainable supply chain management: moving toward new theory. Int J Phys Distrib Logist Manag 38(5):360–387

Cachon GP (2002) Supply coordination with contracts. In: De Kok T, Graves S (eds) Handbooks in operations research and management science. North-Holland, Amsterdam

Cagliano R, Caniato F, Golini R, Kalchschmidt M, Spina G (2008) Supply chain configurations in a global environment: a longitudinal perspective. Oper Manag Res 1:86–94

Caridi M, Cigolini R, De Marco D (2005) Improving supply-chain collaboration by linking intelligent agents to CPFR. Int J Prod Res 43:4191–4218

Chang V (2006) Shimano and the high-end road bike industry, Stanford Graduate School of Business, SM-150

Chen IJ, Paulraj A, Lado AA (2004) Strategic purchasing, supply management, and firm performance. J Oper Manag 22:505–523

Childerhouse P, Aitken J, Towill DR (2002) Analysis and design of focused demand chains. J Oper Manag 20:675–689

Cooper MC, Lambert DM, Pagh PD (1997) Supply chain management: more than a new name for logistics. Int J Logist Manag 8:1–13

Danese P, Romano P (2013) The moderating role of supply network structure on the customer integration-efficiency relationship. Int J Oper Prod Manag 33(4):372–393

Delen D, Benjamin PC (2003) Towards a truly integrated enterprise modeling and analysis environment. Comput Ind 51:257–268

De Matta R, Miller T (2015) Formation of a strategic manufacturing and distribution network with transfer prices. Eur J Oper Res 241:435–448

Fiala P (2005) Information sharing in supply chains. Omega 33:419–423

Fisher M, Hammond J, Obermeyer W, Raman A (1997) Configuring a supply chain to reduce the cost of demand uncertainty. Prod Oper Manag 6:211–225

Fliedner G (2003) CPFR: an emerging supply chain tool. Ind Manag Data Syst 103:14–21

Frohlich MT, Westbrook R (2001) Arcs of integration: an international study of supply chain strategies. J Oper Manag 19:185–200

Galvin P, Morkel A (2001) The effect of product modularity on industry structure: the case of the world bicycle industry. Ind Innovat 8(1):31–47

Gunasekaran A, Lai K, Edwin Cheng TC (2008) Responsive supply chain: a competitive strategy in a networked economy. Omega 36(4):549–564

Jasti NVK, Kodali R (2015) A critical review of lean supply chain management frameworks: proposed framework. Prod Plan Control 26(13):1051–1068

Kliger C, Reuter B, Stadtler H (2015) Collaboration planning. In: Stadtler H, Kliger C, Meyr H (eds) Supply chain management and advanced planning. Springer, New York, pp 257–277

Kobayashi T, Tamaki M, Komoda N (2003) Business process integration as a solution to the implementation of supply chain management systems. Inform Manag 40:769–780

Lau HCW, Lee WB (2000) On a responsive supply chain information system. Int J Phys Distrib Logist 30:598–610

Lee HL (2003) Aligning supply chain strategies with product uncertainties. IEEE Eng Manag Rev 31:26–34

Magal SR, Word J (2012) Integrated business processes with ERP systems. Wiley, New York

Mangiaracina R, Song G, Perego A (2015) Distribution network design: a literature review and a research agenda. Int J Phys Distrib Logist Manag 45(5):506–531

Mentzer JT, DeWitt W, Keebler JS, Min S, Nix NW, Smith CD, Zacharia ZG (2001) Defining supply chain management. J Bus Logist 22:1–25

Meixell MJ, Gargeya VB (2005) Global supply chain design: a literature review and critique. Transport Res E Logist 41:531–550

Ng CK, Vechapikul T (2002) Evaluation of SCOR and CPFR for supply chain collaboration, Proceedings of the Seventh International Conference on Manufacturing and Management, pp 297–304

Sheffi Y (1985) Some analytical problems in logistics research. Transport Res Gen 19:402–405

Simchi-Levi D, Kaminsky P, Simchi-Levi E (2007) Designing and managing the supply chain. McGraw-Hill/Irwin, New York

Soni G, Kodali R (2013) A critical review of supply chain management frameworks: proposed framework. Benchmarking 20(2):263–298

Supply Chain Council (2011) Supply chain operations reference model. Version 10.0

Stadtler H (2008) Supply chain management - an overview. In: Stadtler H, Kliger C (eds) Supply chain management and advanced planning. Springer, New York, pp 9–36

Stewart G (1997) Supply-chain operations reference model (SCOR): the first cross-industry framework for integrated supply-chain management. Logist Inf Manag 10:62–67

Themistocleous M, Iran Z, Love PED (2004) Evaluating the integration of supply chain information systems: a case study. Eur J Oper Res 159:93–405

Verwijmeren M (2004) Software component architecture in supply chain management. Comput Ind 53:165–178

Vonderembse MA, Uppal M, Huang SH, Dismukes JP (2006) Designing supply chains: towards theory development. Int J Prod Econ 100(2):223–238

Chapter 3
Literature Review

3.1 Introduction

Supply chain configuration research has attracted significant attention in scientific literature. This chapter offers a review of these studies and identifies common characteristics of supply chain configuration research. This review is compiled in the form of a table categorizing each paper, considered according to several criteria along with some comments on distinguishing features of the paper. This table can be used as a quick reference for finding papers dealing with the supply chain configuration problem. The chapter also contains summarized results of the complete review.

As described in the previous chapter, supply chain configuration is tightly interrelated with many other supply chain management and general managerial problems. Therefore, some limits in the literature survey are introduced. The survey covers only the core of the supply chain configuration problem without including papers describing general supply chain management methods and technologies important to configuration.

There are several existing surveys on the supply chain configuration problem. Melo et al. (2009) have prepared a comprehensive overview of supply chain configuration and facility location research. They characterize types of supply chain configuration problems and identify decision variables typically considered. Besides location–allocation decisions, capacity, inventory, and production planning decisions are frequently investigated. They identify that single-objective cost minimization dominates and specific algorithms are often developed to solve the configuration problems. Lambiase et al. (2013) analyze existing work to identify strategic decisions, economic parameters, and model features covered by configuration models. Farahani et al. (2015) similarly investigates location-inventory problem including formulation of the basic model. The global aspects of supply chain configuration are summarized by Goetschalckx et al. (2002). These include taxation, cash flow, and trade barriers. The international characteristics trait is

© Springer Science+Business Media New York 2016
C. Chandra, J. Grabis, *Supply Chain Configuration*,
DOI 10.1007/978-1-4939-3557-4_3

further investigated by Meixell and Gargeya (2005). They analyze configuration models according to decision variables (facility location is included in all models considered), performance measures (aftertax profit minimization is important for global models), level of supply chain integration (majority of models consider just two tiers), and globalization considerations. Model complexity characteristics are briefly reviewed by Dasci and Verter (2001). Environmental aspects of supply chain configuration have gained significant attention in recent years. Ramezani et al. (2014) and Das and Rao Posinasetti (2015) among others provide brief focused overview of research in this area. Similarly, Pashaei and Olhager (2015) and Ratha (2014) have explored the topical area of concurrent product and supply chain design. The former review identifies outsourcing, supplier selection, supplier relationships, distance from focal firm, and alignment as key research themes in concurrent design. Bellamy and Basole (2012) review research on supply chain systems and categorize these works pertinent to supply chain structure, dynamics, and strategy. They identify research gaps, which are often associated with supply chain reconfiguration challenges. Decision-making models used in partner selection are analyzed in Wu and Barnes (2011). The review covers models for formulation of criteria, qualification, and final selection.

The following section describes the design of the literature survey, including description of categorization criteria. Sect. 3.3 provides the complete review tables with regard to configuration dimensions and complexity criteria. Results of the review are analyzed in Sect. 3.4, and summary of the chapter is provided in Sect. 3.5.

3.2 The Design of the Literature Survey

The objectives of this state-of-the-art survey are to provide a comprehensive overview of the supply chain configuration problem, to identify main scientific and industrial focus areas, and to quantify the importance of different dimensions of the supply chain configuration problem.

The state-of-the-art review focuses on papers dealing directly with the supply chain configuration. It covers conceptual, model-based, and applied papers to provide a comprehensive overview of different aspects of supply chain configuration. However, it maintains an industrial engineering and computational emphasis.

The main sources of information for the survey are the Scientific Citation Index and Scopus. The main keywords searched for are combinations of "supply chain" or "supply network" with "configuration," "design," and "structure." Some papers found according to these keywords were omitted because they cover issues beyond the scope of definition used in this book. That often occurred often with papers found by using the "design" keyword. Preconditions for including model-based papers in the review are consideration of at least two supply chain tiers and evaluation of multiple alternative supply chain configurations. The second precondition particularly affected inclusion of papers using simulation. Although several papers deal with issues related to strategic supply chain configuration, configuration is often treated as a fixed input parameter without considering any alternatives.

Chronologically, this survey covers the time period from 2007 to 2015. Chandra and Grabis (2007) provided a review for the time period from 1998 to 2006. In Sect. 3.4, results of the current review are often analyzed in comparison with Chandra and Grabis (2007).

The supply chain configuration problem shares many common features with problems, such as distribution planning, supplier selection, manufacturing systems, and facility location. Some references to the most important papers in these areas are included. Readers are referenced to survey papers in these areas for more detailed coverage. Facility location is reviewed by Owen and Daskin (1998) and more recently by Farahani et al. (2015) with focus on the covering problems. De Boer et al. (2001) and Chai et al. (2013) summarize research on supplier selection. Design of manufacturing networks is explored by Shi and Gregory (1998) and Cheng et al. (2015). The total number of papers reviewed in this chapter is 111.

The literature is summarized by classifying papers according to the number of criteria and by evaluating complexity of supply chain configuration problems solved. The following subsections describe these criteria.

3.2.1   Classification Criteria

The literature classification criteria are chosen to represent the most important dimensions of the supply chain configuration problem, as well as describe general characteristics of papers. These dimensions have been identified in previous chapters of the book. Importance and values for each criterion are defined as follows:

3.2.1.1   Horizontal Focus

This criterion describes which tiers of the supply chain are considered in a paper. It allows judging about units assigning the largest value to configuration decisions. Typical values are supply tier, manufacturing tier, distribution tier, and customer tier. In many papers, the whole supply chain is covered, implying that all tiers are under similar levels of consideration.

3.2.1.2   Vertical Focus

It represents location of the problem investigated in the hierarchical decision-making structure comprising strategic, tactical, and operational decision-making levels. The supply chain configuration typically is a strategic problem. However, in order to represent its interactions with other areas of supply chain management, other decision-making levels are also included in decision-making models. Quantification of this criterion allows assessing the importance of each decision-making level.

3.2.1.3   Specific Problem Area

Depending upon supply chain priorities and specific constraints, solving of the supply chain configuration problems can be more tightly coupled with some specific problem areas than others. For instance, inventory management can be of primary concern for supply chains delivering expensive products, while transportation is especially important for global supply chains delivering bulky products.

3.2.1.4   General Problem Area

As with specific problems, a particular general problem (e.g., globalization, coordination) can be the focus of a supply chain configuration study.

3.2.1.5   Modeling Technique

The criterion characterizes a modeling technique used to solve the supply chain configuration problem. Analysis of this criterion reveals the most often used techniques. Values of the criterion include different methods of mathematical programming, simulation, statistical analysis, data modeling, and hybrid techniques. Usually, one method is indicated unless several methods having similar importance to decision-making are used.

3.2.1.6   Application Area

This criterion indicates a particular industry.

3.2.1.7   Type of Paper

This criterion classifies papers as conceptual, model-based, technology, experimental, applied, and survey. Conceptual papers discuss general issues and methodological aspects of the supply chain configuration problem. Model-based papers propose some sort of supply chain configuration models, either quantitative or qualitative. Technology papers develop tools for supply chain configuration decision making or implementation. Extensive numerical studies are provided by papers categorized as experimental. Applied papers focus on solving a particular decision-making problem, and survey papers review existing works on the supply chain configuration.

Not all papers can be classified according to each criterion. For instance, the application area is not defined in all papers.

3.2.2 Complexity Criteria

The papers presenting quantitative models are also evaluated according to several criteria characterizing the complexity of considered supply chain configuration problems. This complexity evaluation is aimed at illustrating what types of problems can be solved in practice. The complexity criteria used in this review are as follows:

Number of units. This substantially influences the complexity of model building (i.e., data gathering is more complex) and the feasibility of model solving. This number generally counts as potential units.

Number of tiers. This influences the complexity of links among supply chain units. Customers are also counted as one supply chain tier.

Persistence. This characterizes whether supply chain configuration is perceived as relatively stable or if models contain some special constructions to represent quickly changing configurations.

Internationalization. Given the fact that many supply chains are multinational, international factors such as tax rates, exchange rates, and duties might have a major impact on configuration decisions. This criteria shows whether international features have been included in the model.

Product variety. Product variety influences the complexity of model development and the feasibility of model solving. This factor is of particular importance because of the increasing role of mass customization.

Integrity. Supply chains generally involve units representing relatively independent units. This criterion indicates whether models treat the supply chain as homogenous, or heterogeneity related issues are addressed.

3.3 Detailed Review

The detailed review is compiled in Tables 3.1 and 3.2, where papers dealing with supply chain configuration are categorized according to configuration dimensions and complexity criteria, respectively. The following abbreviations are used for the classification criteria in Table 3.1: HE—horizontal extent; VE—vertical extent; SP—specific problem; GP—general problem, MT—modeling technique; AA—application area; TP—type of paper. The comments column identifies some of the unique features of the paper.

Table 3.1 Detailed review of supply chain configuration papers

#	Paper	HE	VE	SP	GP	MT	AA	TP	Comments
1	Akanle and Zhang (2008)	S, M	T	PP	RP, CO	AI, GA	EL	QN	Resource configuration options at nodes by agents; clustering to find the best overall configuration
2	Akbari and Karimi (2015)	M, D	S	INV, C		SP		QN, EX	Robust optimization
3	Altiparmak et al. (2009)	A	S			MIP, GA		QN, EX	
4	Altmann (2014)	A	S, T	C, TR, DP	ENV, FIN	MIP	MN	QN, A	Capacity uninstallation cost; financial measures
5	Amaral and Kuettner (2008)	M, D	S	IT			EL	T, A	Spreadsheet based models are used to explore different supply chain configuration scenarios
6	Ameri and McArthur (2013)	S, M	T	SS, IT	CO	AI	MTL	QN, T	
7	Amin and Zhang (2012)	S, M, R	S, T	SS	RL, MO	FZ, MIP	EL	QN	
8	Ashayeri et al. (2012)	A	S, T		MO	FZ		C, QN	Experts define alternative configurations out of SCOR, fuzzy logics is used to evaluate the alternatives
9	Baghalian et al. (2013)	M, D, R	S, T		RE	SP	FD	QN, A	Robust model is developed by considering multiple scenarios
10	Bassett and Gardner (2013)	A	S, T	INV	RE, GL	MIP	CH	QN, A	From generic model to product family models; discussion on application; support for traceability
11	Baud-Lavigne et al. (2014)	A	S	M, BOM, C	ENV, CE, MO	MOP		QN, EX	
12	Bottani and Montanari (2010)	M, D	T, O	INV	DP, CO, RP	SIM	CNM	QN, EX	Information sharing

#	Reference								
13	Brandenburg (2015)	A	S, O	INV		SIM	CNM	QN, A	Value-based approach
14	Brandenburg et al. (2014)	S, M	S, O	INV		SIM	CNM	C, QN, A	Survey based evaluation of different sourcing and distribution strategies
15	Cagliano et al. (2008)	A	S		GL			QL	
16	Castellano et al. (2013)	S, M	S, T	M, BOM, INV	CS	SIM, AI	MN	QN, T	
17	Chaabane et al. (2012)	A, R	S	INV	ENV, RL, ST	MIP	MTL	C, QN, A	Mathematical model is presented as a part of life-cycle approach to sustainable supply chain design
18	Chantarachalee et al. (2014)	S, M	S, O	TR, M, INV	RC	PM, SIM	CRM	QN, A	Lean supply chain
19	Chen (2010)	A	S, T	BOM, C, M, INV	GL, CE	MIP		QN, EX	Item substitution; evaluation of different sourcing strategies
20	Cheung et al. (2012)	S, M	S, O		RP	AI, NA		EX, T	Usage of RFID, case-base reasoning
21	Chiang (2012)	S, M	S		MO, CE	FZ, AHP, DEA, GA	EL	C, QN	Design chain development methodology consisting of three stages; fuzzy hierarchical process
22	Cigolini et al. (2014)	M, D	S, O	INV		SIM		QN, EX	
23	Corominas et al. (2015)	A, R	S, T		MO	NA		C	Supply chain configuration methodology
24	Costa et al. (2010)	A	S			MIP, GA		QN, EX	
25	Costantino et al. (2012)	S, M	S	BOM	MO	MIP		QN, A	Define all possible partners and connects them together; links have different performance attributes; specific structural constraints
26	Creazza et al. (2012a)	D	T	INV, TR		MIP		QN	
27	Creazza et al. (2012b)	D	T	INV, TR, IT		MIP	MV	A	Usage of GIS

(continued)

Table 3.1 (continued)

#	Paper	HE	VE	SP	GP	MT	AA	TP	Comments
28	Cui (2015)	S, M	T	INV, BOM, DP, C	OUT, CE	CP		QN, EX	Outsourcing
29	Das and Rao Posinasetti (2015)	A, R	S		ENV, RL, MO	GP		QN, EX	Impact of design and process selection on environmental
30	de Matta and Miller (2015)	M, D	S		GL	MIP		QN, EX	
31	Diabat et al. (2013)	M, D	S, T	INV		NIP		QN, EX	Usage of Lagrangian relaxation
32	Dubey and Gunasekaran (2016)	D	S		ST			QL	Evaluates alignment, agility, adaptability of different SC designs; humanitarian relief applications
33	ElMaraghy and Mahmoudi (2009)	A	S, T	INV, PP, BOM	GL	MIP	MV	QN	Product modularization scenarios
34	Farahani et al. (2015)	D	S	TR, INV		MOP		S	
35	Fazlollahtabar et al. (2013)	A, R	S, T	INV, TR		FZ, MIP		QN	Comparative study of ranking versus fuzzy mathematical programming
36	Fleuren et al. (2013)	D	S, T	TR	ENV	MIP	3PL	A, QL	Configuration of the express deliveries network; quantification of real-life benefits; emphasis of the user training in using models
37	Ghayebloo et al. (2015)	A, R	S		RL, ENV, MO	MOP		QN, EX	
38	Habermann et al. (2015)	S	S		RE			QN	Empirical evaluation by survey of companies
39	Hearnshaw and Wilson (2013)	A	S		RE, ING	NA		QL	Supply chain characteristics depending on the network measures
40	Hinojosa et al. (2008)	M, D	S, T	INV	RC	MIP		QN	Existing warehouses can be closed
41	Jamshidi et al. (2015)	A	S, T	TR, INV	RP	MIP, GA		QN, EX	
42	Kara and Onut (2010)	R	S		RL	SP	PPR	QN	Uncertain demand and waste data

						NA	MV	QN	
43	Kim et al. (2011)	A	S		RE			QN	Network analysis
44	Klibi and Martel (2012)	D	S	DP, TR		SP		QN, EX	Monte-Carlo for scenario generation; risk evaluation in the second stage
45	Latha Shankar et al. (2013)	A	S			HE		QN, EX	
46	Li and Womer (2008)	A	T	PP		CP	EL	QN	Supply chain configuration is treated as project scheduling
47	Liu and Papageorgiou (2013)	D	S, T	INV	GL, MO	MOP		QN, EX	
48	Longo (2012)	D	S, O	TR	ST, RC	SIM	PHM	QN, A	New locations are represented as experimental factors in the design of experiments
49	Lorentz et al. (2013)	A	S		GL, RC		FD	CS	Defines and analyzes network adjustments depending on context
50	Macchion et al. (2015)	S	S		CO		APL	QL	
51	Madadi et al. (2014)	D	S, T	C	RE	SP	PHM	QN, EX	Two stage model, stochastic factors after facilities are fixed, unreliable supply
52	Maheut et al. (2014)	S, M	T, O	TR, PP, IT		SIM	MN	T, S, A	Decision support system
53	Mansoornejad et al. (2013)	A	S, T	BOM, INV, PP		MIP	BIE	C, QN, A	Methodology, portfolio of network design alternatives, multiple scenarios
54	Medini and Rabénasolo (2014)	D	O		CO	AI		QN, EX	Configuration is defined as scenarios. SCOR model based simulation
55	Meisel and Bierwirth (2014)	A	S, T	TR, PP, BOM		MIP, SIM		QN, EX	Flow can originate at different types of facilities; hybrid simulation and optimization. Simulation uses multiple PP and TR decision-making policies
56	Metta and Badurdeen (2013)	A, R	S		RL, CO			C	

(continued)

Table 3.1 (continued)

#	Paper	HE	VE	SP	GP	MT	AA	TP	Comments
57	Metters and Walton (2007)	D	S				CNM, EL	C	Discussion of differences among different distribution networks in the case of multi-channels
58	Mittermayer and Rodríguez-Monroy (2013)	A	S, O	INV, PP	CO	SIM		QN, EX	
59	Mizgier et al. (2015)	S, M	S		RE	AN		QN, EX	Analyzes impact of hazardous event on different supply chain structures
60	Mohammadi Bidhandi et al. (2009)	A	S			MIP		QN, EX	Lagrangian relaxation
61	Moncayo-Martínez and Zhang (2011)	M, D	S, T		MO	HE	MN	QN	Supply chain is represented as a process where activities can be implemented using different resource options
62	Nasiri et al. (2010)	M, D	S, T, O	INV		MIP		QN, EX	
63	Nepal et al. (2011)	S, M	T, O	INV	MO	GP, GA	MN	QN	At each node supply option is selected
64	Nickel et al. (2012)	D	T		FIN	MIP, SP			Dynamics evaluation of investment decisions and development of scenario tree for representing demand and interest rate uncertainty
65	Olivares-Benitez et al. (2013)	D	S	TR	MO	MOP		QN, EX	
66	Osman and Demirli (2010)	S	T	INV, BOM	OUT	GP	ACF	QN, EX	Computational performance in the case large product variety
67	Ouhimmou et al. (2009)	A	S	INV, PP, IT		MIP	FRN	QN, A, T	Integrated optimization tool; quantification of benefits from using the planning model
68	Paksoy and Özceylan (2013)	A	S	SS		MIP		QN, EX	Supply chain configuration and quality trade-off, contract

#	Reference								
69	Pan and Nagi (2013)	M	T	INV, PP		MIP		QN, EX	
70	Parmigiani et al. (2011)	A	S					QL	The supply chain configuration drives the development of distinctive portfolios of social and environmental capabilities
71	Paydar and Saidi-Mehrabad (2015)	S, M	S, O	PP, INV	MO	SP	MN	QN, A	Supply chain is optimized simultaneously with cell formation
72	Petridis (2015)	M, D	S	INV	CS	NIP		QN, EX	(Q,R) inventory control policy is included in the configuration model
73	Pirard et al. (2011)	A	O	INV, PP		SIM	PHM, PPR	QN, EX	This rule aims to allocate a demand expressed by a customer to a distribution site
74	Pokharel (2008)	A	S		MO, GL	MOP		QN, EX	The approach presented here is to facilitate the decision-making process and not to force the decision
75	Prakash et al. (2012)	S, D	S	TR	MO	GA	APL	QN, EX	
76	Qu et al. (2010a)	S	T, O	IT	ING, CO, RC		EL	T	atcPortal, a generic portal for supply chain. Web services for decentralized decision-making
77	Qu et al. (2010b)	S	S		CO	MIP, GA	EL	QN, EX	Analytical target cascading (ATC) for configuring assembly supply chains with convergent structures. Individual enterprises in a supply chain are represented as separate elements in an ATC hierarchy. They are able to maintain autonomous and heterogeneous decision systems for optimizing their private decision variables and objectives
78	Qu et al. (2015)	S, M	S		CO	MIP		QN, EX	Supply chain formation in clusters; different order allocation policies for a leader; design problem is divided in sub-problems
79	Quariguasi Frota Neto et al. (2008)	A, R	S		RL, MO, ST	MOP, DEA	PPR	QN	Framework for a sustainable logistic network

(continued)

Table 3.1 (continued)

#	Paper	HE	VE	SP	GP	MT	AA	TP	Comments
80	Ramezani et al. (2013)	A, R	S	C	MO, RL	MOP, SP		QN, EX	Combines forward/reverse logistics
81	Ramezani et al. (2014)	A, R	S, T	INV	FIN, RL	MIP		QN, EX	Closed-loop network with repair, remanufacturing, and disposal options
82	Rigot-Muller et al. (2013)	D	S, T	TR	ENV	PM	3PL	QN, A	Value stream mapping
83	Roni et al. (2014)	S, M	S	T		MIP	BIE	QN, EX	Benders' decomposition. Hub-and-spoke supply network design
84	Santibanez-Gonzalez and Diabat (2013)	R	S		RL	MIP		QN, EX	Benders' decomposition
85	Scott et al. (2015)	S	S, T	SS	MO	AHP, SIM	BIE	C, QN, A	Modeling methodology including the conceptual modeling; multiple stakeholders
86	Selim and Ozkarahan (2008)	M, D	S	C	MO	GP, FZ		QN, EX	
87	Shabani and Sowlati (2013)	S, M	T	INV		NIP	BIE	QN, A	
88	Shahzad and Hadj-Hamou (2013)	A	S, T	INV, BOM		MIP	EL	QN, EX	Benders' decomposition
89	Soleimani et al. (2013)	A, R	S, T	INV	RL	MIP, GA		QM, EX	Network of reverse logistics
90	Sousa et al. (2008)	M, D	T, O	INV, TR	GL	MIP	CH	QN, A	Two-stage methodology to asses feasibility of plans at the operational level
91	Srai and Gregory (2008)	A	S		GL			C, CS	
92	Su et al. (2012)	S, M	T			DP	FRN	QN, A	
93	Tian and Yue (2014)	A	S	C	CS	SP		QN, A, EX	Multiple scenarios are used

#	Author	S, M	S	TR	CS	SIM	BIE	QN, A	
94	Tiacci et al. (2014)	S, M	S			MIP, HE		QN, A	
95	Tiwari et al. (2012)	A	S			AN		QN, EX	
96	Ülkü and Schmidt (2011)	S, M	S		CE, CO	AN		QN, EX	Link between product architecture and supply chain configuration
97	Vahdani and Naderi-Beni (2014)	M, D, R	S	TR	RL	MIP, SP	MTL	QN, EX	Interval programming to account for uncertainty in parameters
98	Vanteddu et al. (2011)	S	T	INV, SS	RP	AN	MV	QN, A	Alternative suppliers are evaluated using analytical expressions
99	Verdouw et al. (2011)	M, D	S, T, O	INV, TR, M	CO	PM	CNM	C, T	Manufacturing policies, SCOR based process design
100	von Massow and Canbolat (2014)	A	S			MIP		C, QN	Performance score imposes meeting the characteristic score requirements
101	Wang et al. (2011)	A	S		ENV, MO	MIP		QN, EX	
102	Wilhelm et al. (2013)	A	S, T	C	RC	MIP		QN, EX	Network-based model for reconfiguration
103	Wu and Zhang (2014)	S, D	S	INV, DP		NIP		QN, EX	Demand uncertainty
104	Yadav et al. (2009)	A	T			MIP, HE	EL	QN, EX	Algorithm portfolio
105	Yang et al. (2015)	A	S, T	BOM	CE, CO	MIP, GA	MN	QN, A	Coordination between product designers and supply chain managers following the Stackelberg game approach
106	Yao et al. (2010)	A	T	INV, PP		MIP		QN, EX	
107	You and Grossmann (2008)	A	S, T	INV, C	RP, CS	NIP	CH	QN, A	
108	Zhang et al. (2008)	S, M	S, T	INV	CE, OUT	MIP, AN	EL	QN, EX	Product variability model combined with supply chain topology

(continued)

Table 3.1 (continued)

#	Paper	HE	VE	SP	GP	MT	AA	TP	Comments
109	Zhang et al. (2010)	D	O	INV	CO	AI	APL	QN, EX	
110	Zhang et al. (2013)	S, M	S, T	BOM, M	GL, MO	MIP	APL	QN, A	
111	Zokaee et al. (2014)	A	S		CS	SP	FD	QN, A	

Key:

HE (Horizontal Focus): *A* all supply chain tiers (not including return), *D* distribution tier, *M* manufacturing tier, *S* supply tier, *R* return

VE (Vertical Focus): *S* strategic decision-making, *T* tactical decision-making, *O* operational decision-making

SP (Specific Problem): *BOM* bill-of-materials, *C* capacity, *DP* demand planning, *INV* inventory, *IT* information technology, *PP* production planning and manufacturing , *SS* supplier selection, *TR* transportation

GP (General Problem): *CE* concurrent engineering, *CO* coordination, *CS* customer service, *ENV* environmental protection, *FIN* finance, *GL* globalization, *ING* integration, *MO* multiple objectives, *OUT* outsourcing, *RC* reconfigurability, *RE* reliability, resilience, robustness, and traceability, *RL* reverse logistics, *RP* responsiveness, *ST* sustainability

MT (Modeling Technique): *AI* artificial intelligence and agent, *AN* analytical, *AHP* analytic-hierarchical process, *CP* constraint programming, *DEA* data envelopment analysis, *DP* dynamic programming, *FZ* fuzzy modeling, *GA* genetic algorithms, *GP* goal programming, *HE* heuristic, *MIP* mixed integer programming, *MOP* multi-programming, *NA* network analysis and graph theory, *NIP* nonlinear programming, *PM* process modeling, *SIM* simulation, *SP* stochastic programming

AA (Application Area): *3PL* third party logistics and maritime, *ACF* Aircraft, *APL* Apparel, *BIE* bioenergy and energy, *CH* chemicals, *CNM* consumer goods, *CRM* construction materials, *EL* electronics, *FD* food, *FRN* furniture, *MN* machinery and equipment, *MTL* metal, *MV* motor vehicles, *PHM* pharmaceutical, *PPR* pulp and paper

TP (Type of Paper): *A* applied, *C* conceptual, *CS* case study, *EX* experimental, *QL* qualitative, *QN* quantitative, *S* survey, *T* technology

Table 3.2 Detailed evaluation of complexity of selected supply chain problems

Paper	Number of units	Number of tiers	Time horizon	Globalization	Product variety	Integrity
Altiparmak et al. (2009)	<1000	4	Single	–	Low	High
Baghalian et al. (2013)	<100	3	Single	Local	Low	High
Bottani and Montanari (2010)	<10	3	Multiple	Local	Low	Medium
Castellano et al. (2013)	<100	3	Multiple	Europe	Medium	High
Chaabane et al. (2012)	<10	4	Multiple	Local	Low	High
Chen (2010)	<10	3	Single	Global	Medium	Medium
Kara and Onut (2010)	<100	3	Single	Local, Turkey	Low	High
Klibi and Martel (2012)	<1000	2	Multiple	US	Low	High
Liu and Papageorgiou (2013)	<100	2	Multiple	Global	Medium	High
Meisel and Bierwirth (2014)	<100	4	Multiple	-	Medium	High
Nasiri et al. (2010)	<1000	3	Multiple	Local	Low	High
Paydar and Saidi-Mehrabad (2015)	<10	3	Multiple	Iran	Medium	High
Pirard et al. (2011)	<100	4	Multiple	Europe	Medium	Medium
Roni et al. (2014)	<1000	3	Singe	US	Low	High
Shabani and Sowlati (2013)	<100	3	Multiple	Canada	Low	High
Shahzad and Hadj-Hamou (2013)	<100	3	Multiple	-	High	High
Soleimani et al. (2013)	<200	6+	Multiple	–	Low	High
Vanteddu et al. (2011)	<10	2	Multiple	US	Low	High
Wang et al. (2011)	<100	3	Single	China, Global	Low	High

3.4 Focus Areas for Supply Chain Configuration

Results of the detailed review are cross-tabulated to identify focus areas of supply chain configuration research. Tables 3.3 and 3.4 report the cross-tabulation results according to horizontal and vertical extent dimensions, respectively. A majority of papers attempt to cover all supply chain tiers. Some of the papers focus on the supply or distribution sites having the manufacturing tier as an integrative stage. However, there are differences between papers focusing on the distribution and

Table 3.3 Number of papers according to the horizontal extent (HE) dimension

Value	Whole supply chain	Distribution	Supply	Manufacturing	Return
Number of papers	47	30	32	38	15

Table 3.4 Number of papers according to the vertical extent (VE) dimension

Value	Strategic	Tactical	Operational
Number of papers	90	48	19

supply tiers as location decisions mainly occur in the distribution and manufacturing tiers, while decision made in the supply tier are limited mainly to allocation and selection. There is a significant increase of papers including the return tier. For instance, Soleimani et al. (2013) investigates a complete reverse chain spanning from distribution to supply tiers. Although the supply chain configuration decisions are primarily perceived as strategic decisions, there is a significant increase of papers incorporating tactical and operational aspects what allows for joint evaluation of supply chain configuration decisions and other operations management decisions. Papers incorporating tactical aspects often focus on inventory management and quarterly planning.

The results for the vertical focus dimension relate to most often considered specific problems (see Table 3.5). The table reports only those specific problems that have been addressed in more than one paper. The results suggested that typical supply chain configuration decisions of location, selection, and allocation are increasingly combined with other aspects of supply chain and operations management and models are becoming more complex and elaborate. Inventory management is the most often considered specific problem. Usually, it is addressed in multi-period models, and safety stock requirements are also included in some models. While transportation flows are present in almost any configuration model, more detailed representation of the transportation problem is in 19 papers what includes nonlinear transportation costs, detailed choice of transportation mode, and analysis of transit time. Product design issues are often addressed concurrently with supply chain design and the bill-of-material is incorporated in 11 of the reviewed papers. The demand planning papers specifically account for demand volatility and its impact on configuration decisions. Issues related to information technology are generally investigated in papers exclusively devoted to this problem and, therefore, less frequently addressed with explicitly focusing on supply chain configuration.

Balancing of multiple objectives is the most often considered general problem (see Table 3.6). That is especially important in papers also focusing on supply chain responsiveness and environmental factors. Accounting for multiple objectives is achieved using various modeling techniques including multi-objective programming and simulation. The relative importance of the globalization aspects has decreased while the number of papers devoted to environmental aspects and reverse

Table 3.5 Papers considering particular specific problems (SP)

Specific problem	Paper
BOM (12)	Baud-Lavigne et al. (2014), Castellano et al. (2013), Chen (2010), Costantino et al. (2012), Cui (2015), ElMaraghy and Mahmoudi (2009), Mansoornejad et al. (2013), Meisel and Bierwirth (2014), Osman and Demirli (2010), Shahzad and Hadj-Hamou (2013), Yang et al. (2015), Zhang et al. (2013)
C (11)	Akbari and Karimi (2015), Altmann (2014), Baud-Lavigne et al. (2014), Chen (2010), Cui (2015), Madadi et al. (2014), Ramezani et al. (2013), Selim and Ozkarahan (2008), Tian and Yue (2014), Wilhelm et al. (2013), You and Grossmann (2008)
DP (5)	Altmann (2014), Brandenburg (2015), Cui (2015), Klibi and Martel (2012), Yadav et al. (2009)
INV (42)	Akbari and Karimi (2015), Bassett and Gardner (2013), Bottani and Montanari (2010), Brandenburg (2015), Brandenburg et al. (2014), Castellano et al. (2013), Chaabane et al. (2012), Chantarachalee et al. (2014), Chen (2010), Cigolini et al. (2014), Creazza et al. (2012a), Creazza et al. (2012b), Cui (2015), Diabat et al. (2013), ElMaraghy and Mahmoudi (2009), Farahani et al. (2015), Fazlollahtabar et al. (2013), Hinojosa et al. (2008), Jamshidi et al. (2015), Liu and Papageorgiou (2013), Mansoornejad et al. (2013), Mittermayer and Rodríguez-Monroy (2013), Nasiri et al. (2010), Nepal et al. (2011), Osman and Demirli (2010), Ouhimmou et al. (2009), Pan and Nagi (2013), Paydar and Saidi-Mehrabad (2015), Petridis (2015), Pirard et al. (2011), Ramezani et al. (2014), Shabani and Sowlati (2013), Shahzad and Hadj-Hamou (2013), Soleimani et al. (2013), Sousa et al. (2008), Vanteddu et al. (2011) , Verdouw et al. (2011), Yadav et al. (2009), Yao et al. (2010), You and Grossmann (2008), Zhang et al. (2008), Zhang et al. (2010)
IT (6)	Amaral and Kuettner (2008), Ameri and McArthur (2013), Creazza et al. (2012b), Maheut et al. (2014), Ouhimmou et al. (2009), Qu et al. (2010a)
PP (19)	Akanle and Zhang (2008), Baud-Lavigne et al. (2014), Brandenburg (2015), Castellano et al. (2013), Chantarachalee et al. (2014), Chen (2010), ElMaraghy and Mahmoudi (2009), Li and Womer (2008), Maheut et al. (2014), Mansoornejad et al. (2013), Meisel and Bierwirth (2014), Mittermayer and Rodríguez-Monroy (2013), Ouhimmou et al. (2009), Pan and Nagi (2013), Paydar and Saidi-Mehrabad (2015), Pirard et al. (2011), Verdouw et al. (2011), Yao et al. (2010), Zhang et al. (2010)
SS (5)	Ameri and McArthur (2013), Amin and Zhang (2012), Paksoy and Özceylan (2013), Scott et al. (2015), Vanteddu et al. (2011)
TR (19)	Altmann (2014), Brandenburg (2015), Chantarachalee et al. (2014), Creazza et al. (2012a), Creazza et al. (2012b), Farahani et al. (2015), Fazlollahtabar et al. (2013), Fleuren et al. (2013), Jamshidi et al. (2015), Klibi and Martel (2012), Longo (2012), Maheut et al. (2014), Meisel and Bierwirth (2014), Olivares-Benitez et al. (2013), Prakash et al. (2012), Rigot-Muller et al. (2013), Sousa et al. (2008), Tiacci et al. (2014), Vahdani and Naderi-Beni (2014), Verdouw et al. (2011)

Note: The number of papers for each specific problem is given in parenthesis. See key from Table 3.1 for abbreviations

Table 3.6 Papers considering particular general problems (GP)

General problem	Paper
CE (8)	Baud-Lavigne et al. (2014), Brandenburg (2015), Chen (2010), Chiang (2012), Cui (2015), Ülkü and Schmidt (2011), Yang et al. (2015), Zhang et al. (2008)
CO (14)	Akanle and Zhang (2008), Ameri and McArthur (2013), Bottani and Montanari (2010), Macchion et al. (2015), Medini and Rabénasolo (2014), Metta and Badurdeen (2013), Mittermayer and Rodríguez-Monroy (2013), Qu et al. (2010a), Qu et al. (2010b), Qu et al. (2015), Ülkü and Schmidt (2011), Verdouw et al. (2011), Yang et al. (2015), Zhang et al. (2010)
CS (7)	Brandenburg (2015), Castellano et al. (2013), Petridis (2015), Tian and Yue (2014), Tiacci et al. (2014), You and Grossmann (2008), Zokaee et al. (2014)
ENV (9)	Altmann (2014), Baud-Lavigne et al. (2014), Brandenburg (2015), Chaabane et al.(2012), Das and Rao Posinasetti (2015), Fleuren et al. (2013), Ghayebloo et al. (2015), Rigot-Muller ct al. (2013), Wang et al. (2011)
FIN (4)	Altmann (2014), Brandenburg (2015), Nickel et al. (2012), Ramezani et al. (2014)
GL (11)	Bassett and Gardner 2013, Cagliano et al. (2008), Chen (2010), de Matta and Miller (2015), ElMaraghy and Mahmoudi (2009), Liu and Papageorgiou (2013), Lorentz et al. (2013), Pokharel (2008), Sousa et al. (2008), Srai and Gregory (2008), Zhang et al. (2013)
ING (2)	Hearnshaw and Wilson (2013), Qu et al. (2010a)
MO (22)	Amin and Zhang (2012), Ashayeri et al. (2012), Baud-Lavigne et al. (2014), Brandenburg (2015), Chiang (2012), Corominas et al. (2015), Costantino et al. (2012), Das and Rao Posinasetti (2015), Ghayebloo et al. (2015), Liu and Papageorgiou (2013), Moncayo-Martínez and Zhang (2011), Nepal et al. (2011), Olivares-Benitez et al. (2013), Paydar and Saidi-Mehrabad (2015), Pokharel (2008), Prakash et al. (2012), Quariguasi Frota Neto et al. (2008), Ramezani et al. (2013), Scott et al. (2015), Selim and Ozkarahan (2008), Wang et al. (2011), Zhang et al. (2013)
OUT (3)	Cui (2015), Osman and Demirli (2010), Zhang et al. (2008)
RC (6)	Chantarachalee et al. (2014), Hinojosa et al. (2008), Longo (2012), Lorentz et al. (2013), Qu et al. (2010a), Wilhelm et al. (2013)
RE (7)	Baghalian et al. (2013), Bassett and Gardner (2013), Habermann et al. (2015), Hearnshaw and Wilson (2013), Klibi and Martel (2012), Madadi et al. (2014), Mizgier et al. (2015)
RL (12)	Amin and Zhang (2012), Chaabane et al. (2012), Das and Rao Posinasetti (2015), Ghayebloo et al. (2015), Kara and Onut (2010), Metta and Badurdeen (2013), Quariguasi Frota Neto et al. (2008), Ramezani et al. (2013), Ramezani et al. (2014), Santibanez-Gonzalez and Diabat (2013), Soleimani et al. (2013), Vahdani and Naderi-Beni (2014)
RP (6)	Akanle and Zhang (2008), Bottani and Montanari (2010), Cheung et al. (2012), Jamshidi et al. (2015), Vanteddu et al. (2011), You and Grossmann (2008)
ST (4)	Chaabane et al. (2012), Dubey and Gunasekaran (2016), Longo (2012), Quariguasi Frota Neto et al. (2008)

Note: The number of papers for each general problem is given in parenthesis. See key from Table 3.1 for abbreviations logistics is increasing. Reliability and sustainability are also the trending supply chain configuration research areas. The reliability research includes a range of papers addressing supply chain resilience to random shocks, robustness as well as traceability (Bassett and Gardner 2013). The sustainability research goes beyond environmental sustainability and also addresses factors related to social and cultural sustainability.

In the context of this book which advocates the importance of reconfigurable supply chains, there are just seven papers that explicitly address the problem of dynamic supply chain reconfiguration.

Mixed-integer programming is the most often used modeling technique (Table 3.7). This observation is consistent with conclusions drawn by Wu and Barnes (2011) even though they focus on partner/supplier selection where methods

Table 3.7 Papers using particular modeling techniques (MT)

Modeling technique	Paper
AI (6)	Akanle and Zhang (2008), Ameri and McArthur (2013), Castellano et al. (2013), Cheung et al. (2012), Medini and Rabénasolo (2014), Zhang et al. (2010)
AN (15)	Cagliano et al. (2008), Dubey and Gunasekaran (2016), Habermann et al. (2015), Kim et al. (2011), Lorentz et al. (2013), Macchion et al. (2015), Metta and Badurdeen (2013), Metters and Walton (2007), Mizgier et al. (2015), Parmigiani et al. (2011), Qu et al. (2010a), Srai and Gregory (2008), Ülkü and Schmidt (2011), Vanteddu et al. (2011), Zhang et al. (2008)
AHP (2)	Chiang (2012), Scott et al. (2015)
CP (2)	Cui (2015), Li and Womer (2008)
DEA (2)	Chiang (2012), Quariguasi Frota Neto et al. (2008)
DP (1)	Su et al. (2012)
GA (10)	Akanle and Zhang (2008), Altiparmak et al. (2009), Chiang (2012), Costa et al. (2010), Jamshidi et al. (2015), Nepal et al. (2011), Prakash et al. (2012), Qu et al. (2010b), Soleimani et al. (2013), Yang et al. (2015)
GP (5)	Brandenburg (2015), Das and Rao Posinasetti (2015), Nepal et al. (2011), Osman and Demirli (2010), Selim and Ozkarahan (2008)
HE (4)	Latha Shankar et al. (2013), Moncayo-Martínez and Zhang (2011), Tiwari et al. (2012), Yadav et al. (2009)
MIP (42)	Altiparmak et al. (2009), Altmann (2014), Amin and Zhang (2012), Bassett and Gardner (2013), Chaabane et al. (2012), Chen (2010), Costa et al. (2010), Costantino et al. (2012), Creazza et al. (2012a), Creazza et al. (2012b), de Matta and Miller (2015), ElMaraghy and Mahmoudi (2009), Fazlollahtabar et al. (2013), Fleuren et al. (2013), Hinojosa et al. (2008), Jamshidi et al. (2015), Mansoornejad et al. (2013), Meisel and Bierwirth (2014), Mohammadi Bidhandi et al. (2009), Nasiri et al. (2010), Nickel et al. (2012), Ouhimmou et al. (2009), Paksoy and Özceylan (2013), Pan and Nagi (2013), Qu et al. (2010b), Qu et al. (2015), Ramezani et al. (2014), Roni et al. (2014),

(continued)

Table 3.7 (continued)

Modeling technique	Paper
	Santibanez-Gonzalez and Diabat (2013), Shahzad and Hadj-Hamou (2013), Soleimani et al. (2013), Sousa et al. (2008), Tiwari et al. (2012), Vahdani and Naderi-Beni (2014), von Massow and Canbolat (2014), Wang et al. (2011), Wilhelm et al. (2013), Yadav et al. (2009), Yang et al. (2015), Yao et al. (2010), Zhang et al. (2008), Zhang et al. (2013)
MOP (8)	Baud-Lavigne et al. (2014), Farahani et al. (2015), Ghayebloo et al. (2015), Liu and Papageorgiou (2013), Olivares-Benitez et al. (2013), Pokharel (2008), Quariguasi Frota Neto et al. (2008), Ramezani et al. (2013)
NA (4)	Cheung et al. (2012), Corominas et al. (2015), Hearnshaw and Wilson (2013), Kim et al. (2011)
NIP (5)	Diabat et al. (2013), Petridis (2015), Shabani and Sowlati (2013), Wu and Zhang (2014), You and Grossmann (2008)
PM (3)	Chantarachalee et al. (2014), Rigot-Muller et al. (2013), Verdouw et al. (2011)
SIM (12)	Bottani and Montanari (2010), Brandenburg et al. (2014), Castellano et al. (2013), Chantarachalee et al. (2014), Cigolini et al. (2014), Longo (2012), Maheut et al. (2014), Meisel and Bierwirth (2014), Mittermayer and Rodríguez-Monroy (2013), Pirard et al. (2011), Scott et al. (2015), Tiacci et al. (2014)
SP (11)	Akbari and Karimi (2015), Baghalian et al. (2013), Kara and Onut (2010), Klibi and Martel (2012), Madadi et al. (2014), Nickel et al. (2012), Paydar and Saidi-Mehrabad (2015), Ramezani et al. (2013), Tian and Yue (2014), Vahdani and Naderi-Beni (2014), Zokaee et al. (2014)
FZ (5)	Amin and Zhang (2012), Ashayeri et al. (2012), Chiang (2012), Fazlollahtabar et al. (2013), Selim and Ozkarahan (2008)

Note: The number of papers for each modeling technique is given in parenthesis

like AHP and fuzzy sets theory are often used. Analytical techniques are mainly used for comparison of multiple supply chain strategies rather than directly for location decisions as well as for inventory management decisions. Simulation is the third most often used technique. However, given that simulation models have wider scope than mathematical programming models, it is often difficult to decide on categorization of simulation models. Genetic algorithms are usually used to solve these models. A variety of mathematical programming techniques and heuristic methods are used to deal with the increasing complexity of the configuration models. Network analysis methods have found application in evaluation of supply chain configuration strategies and large scale networks.

Table 3.8 lists papers reporting applications in particular industries. The electronics industry is most often considered. Configuration of computer manufacturing supply chains is a particularly popular application case. That is partially explained by many papers using the supply chain example by Graves and Willems (2005) as a benchmark case. The bioenergy production supply chains are investigated in four papers. That is because of socioeconomic importance of the topic and high transportation costs requiring efficient location of supply chain units. However, majority

Table 3.8 Papers reporting application in a particular industry

Industry	Paper
Aircraft (1)	Osman and Demirli (2010)
Bioenergy and energy (5)	Mansoornejad et al. (2013), Roni et al. (2014), Scott et al. (2015), Shabani and Sowlati (2013), Tiacci et al. (2014)
Chemicals (3)	Bassett and Gardner (2013), Sousa et al. (2008), You and Grossmann (2008)
Consumer goods (5)	Bottani and Montanari (2010), Brandenburg (2015), Brandenburg et al. (2014), Metters and Walton (2007) Verdouw et al. (2011)
Construction materials (1)	Chantarachalee et al. (2014)
Electronics (7)	Akanle and Zhang (2008), Amaral and Kuettner (2008), Amin and Zhang (2012), Qu et al. (2010a), Qu et al. (2010b), Shahzad and Hadj-Hamou (2013), Yadav et al. (2009)
Food (3)	Baghalian et al. (2013), Lorentz et al. (2013), Zokaee et al. (2014)
Furniture (2)	Ouhimmou et al. (2009), Su et al. (2012)
Machinery and equipment (2)	Altmann (2014), Yang et al. (2015)
3PL and Maritime (2)	Fleuren et al. (2013), Rigot-Muller et al. (2013)
Metals (3)	Ameri and McArthur (2013), Chaabane et al. (2012), Vahdani and Naderi-Beni (2014)
Motor vehicles (4)	Creazza et al. (2012b), ElMaraghy and Mahmoudi (2009), Kim et al. (2011), Vanteddu et al. (2011)
Pharmaceutical (3)	Longo (2012), Madadi et al. (2014), Pirard et al. (2011)
Pulp and paper (3)	Kara and Onut (2010), Pirard et al. (2011), Quariguasi Frota Neto et al. (2008),
Textiles and apparel (3)	Macchion et al. (2015), Prakash et al. (2012), Zhang et al. (2013)

Note: The number of papers for each industry is given in parenthesis

Table 3.9 The number of papers according to their type

	A	C	CS	EX	QL	QN	S	T
Number of papers	30	9	2	51	6	90	4	8

Note: Types are abbreviated as: *A* applied, *C* conceptual, *CS* case study, *EX* experimental, *QL* qualitative, *QN* quantitative, *S* survey, *T* technology

of applications are reported as examples and there are few papers focusing on actual gains from supply chain configuration. Bassett and Gardner (2013) is one of the papers discussing implementation of configuration decisions and evaluation of post-implementation performance.

Finally, categorization of papers according to type of paper (see Table 3.9) shows that the majority of papers are devoted to quantitative modeling. There are a fair number of papers addressing conceptual issues of supply chain configuration. However, that is often done in an informal manner, which is also confirmed by the small number of model-based qualitative papers. Not all papers focusing on particular application areas (Table 3.8) actually apply their models in industrial cases. Applications are reported in 25% of the papers, which is a significant increase relative to 10% in Chandra and Grabis (2007).

Analysis of the review results according to the problem complexity criteria shows that solving relatively large and complex problems is possible (Table 3.2). Six out of the reviewed papers considered supply chains between 100 and 1000 units (Klibi and Martel 2012, Nasiri et al. 2010, Altiparmak et al. 2009, Costa et al. 2010, Roni et al. 2014, Santibanez-Gonzalez and Diabat 2013). However, one notable observation is that larger problems are usually solved in papers explicitly devoted to developing efficient model-solving algorithms, while papers oriented toward applications and expanding modeling scope usually treat problems of smaller sizes. The large-scale models also are usually single-period models with the low level of product variety.

The majority of papers either do not address the internationalization problem, or they indicate that they deal with local problems. One of the key limitations of existing models is low product variety which persists regardless of recent attention to concurrent engineering. The existing models also tend to represent supply chains as relatively homogeneous entities.

3.5 Summary

One hundred and eleven papers have been identified as dealing directly with the supply chain configuration problem. These papers are categorized according to supply chain configuration dimensions and problem complexity criteria. The list of papers is representative, though we cannot claim complete coverage. Empirical and survey type of papers exploring current practices rather than focusing directly on configuration decision-making are particularly underrepresented.

The literature review suggests that there are several emerging areas of supply chain configuration research, such as:

- Concurrent supply chain, process, and product design
- Increasing importance of inventory management issues in supply chain configuration and fusion between strategic and tactical decision-making
- Multi-objective supply chain optimization, where responsiveness is one of common considerations and resilience and sustainability is gaining increasing attention

At the same time, important issues, such as selection of transportation options and supply chain power structure have attained limited exposure in current literature. Additionally, models tend to address only the general or specific problems. For instance, coordination and integration are usually investigated in relation to the information technology problem, while they are investigated together with inventory management to a limited extent.

References

Akanle OM, Zhang DZ (2008) Agent-based model for optimising supply-chain configurations. Int J Prod Econ 115:444–460

Akbari AA, Karimi B (2015) A new robust optimization approach for integrated multi-echelon, multi-product, multi-period supply chain network design under process uncertainty. Int J Adv Manuf Technol 79:229–244

Altiparmak F, Gen M, Lin L, Karaoglan I (2009) A steady-state genetic algorithm for multi-product supply chain network design. Comput Ind Eng 56:521–537

Altmann M (2014) A supply chain design approach considering environmentally sensitive customers: the case of a German manufacturing SME. Int J Prod Res 53:1–17

Amaral J, Kuettner D (2008) Analyzing supply chains at HP using spreadsheet models. Interfaces 38(4):228–240

Ameri F, McArthur C (2013) A multi-agent system for autonomous supply chain configuration. Int J Adv Manuf Technol 66:1097–1112

Amin SH, Zhang G (2012) An integrated model for closed-loop supply chain configuration and supplier selection: Multi-objective approach. Expert Syst Appl 39:6782–6791

Ashayeri J, Tuzkaya G, Tuzkaya UR (2012) Supply chain partners and configuration selection: an intuitionistic fuzzy Choquet integral operator based approach. Expert Syst Appl 39:3642–3649

Baghalian A, Rezapour S, Farahani RZ (2013) Robust supply chain network design with service level against disruptions and demand uncertainties: a real-life case. Eur J Oper Res 227:199–215

Bassett M, Gardner L (2013) Designing optimal global supply chains at Dow AgroSciences. Ann Oper Res 203:187–216

Baud-Lavigne B, Agard B, Penz B (2014) Environmental constraints in joint product and supply chain design optimization. Comput Ind Eng 76:16–22

Bellamy MA, Basole RC (2012) Network analysis of supply chain systems: a systematic review and future research. Syst Eng 16:235–249

Bottani E, Montanari R (2010) Supply chain design and cost analysis through simulation. Int J Prod Res 48:2859–2886

Brandenburg M (2015) Low carbon supply chain configuration for a new product – a goal programming approach. Int J Prod Res 53:1–23

Brandenburg M, Kuhn H, Schilling R, Seuring S (2014) Performance- and value-oriented decision support for supply chain configuration. Logist Res 7:118

Cagliano R, Caniato F, Golini R, Kalchschmidt M, Spina G (2008) Supply chain configurations in a global environment: a longitudinal perspective. Oper Manag Res 1:86–94

Castellano E, Besga JM, Uribetxebarria J, Saiz E (2013) Supply network configuration. In: Poler R et al (eds) Intelligent non-hierarchical manufacturing networks. Wiley, Hoboken, pp 73–105

Chaabane A, Ramudhin A, Paquet M (2012) Design of sustainable supply chains under the emission trading scheme. Int J Prod Econ 135:37–49

Chandra C, Grabis J (2007) Supply chain configuration. Springer, New York

Chai J, Liu JNK, Ngai EWT (2013) Application of decision-making techniques in supplier selection: A systematic review of literature. Expert Syst Appl 40(10):3872–3885

Chantarachalee K, Carvalho H, Cruz-Machado VA (2014) Proceedings of the eighth international conference on management science and engineering management. Adv Intell Syst Comput 280:797–807

Chen HY (2010) The impact of item substitutions on production-distribution networks for supply chains. Transp Res Part E Logist Transp Rev 46:803–819

Cheng Y, Farooq S, Johansen J (2015) International manufacturing network: Past, present, and future. Int J Oper Prod Manag 3:392–429

Cheung CF, Cheung CM, Kwok SK (2012) A knowledge-based customization system for supply chain integration. Expert Syst Appl 39:3906–3924

Chiang TA (2012) Multi-objective decision-making methodology to create an optimal design chain partner combination. Comput Ind Eng 63:875–889

Cigolini R, Pero M, Rossi T, Sianesi A (2014) Linking supply chain configuration to supply chain performance: a discrete event simulation model. Simul Model Pract Theory 40:1–11

Corominas A, Mateo M, Ribas I, Rubio S (2015) Methodological elements of supply chain design. Int J Prod Res 53:1–14

Costa A, Celano G, Fichera S, Trovato E (2010) A new efficient encoding/decoding procedure for the design of a supply chain network with genetic algorithms. Comput Ind Eng 59:986–999

Costantino N, Dotoli M, Falagario M, Fanti MP, Mangini AM (2012) A model for supply management of agile manufacturing supply chains. Int J Prod Econ 135:451–457

Creazza A, Dallari F, Rossi T (2012a) An integrated model for designing and optimising an international logistics network. Int J Prod Res 50:2925–2939

Creazza A, Dallari F, Rossi T (2012b) Applying an integrated logistics network design and optimisation model: the Pirelli Tyre case. Int J Prod Res 50:3021–3038

Cui LX (2015) Towards optimal configuration of a manufacturer's supply network with demand flexibility. Int J Prod Res 53:3541–3560

Das K, Rao Posinasetti N (2015) Addressing environmental concerns in closed loop supply chain design and planning. Int J Prod Econ 163:34–47

Dasci A, Verter V (2001) A continuous model for production-distribution system design. Eur J Oper Res 129:287–298

De Boer L, Labro E, Morlacchi P (2001) A review of methods supporting supplier selection. Eur J Purchasing Supply Manage 7:75–89

De Matta R, Miller T (2015) Formation of a strategic manufacturing and distribution network with transfer prices. Eur J Oper Res 241:435–448

Diabat A, Richard J-P, Codrington CW (2013) A Lagrangian relaxation approach to simultaneous strategic and tactical planning in supply chain design. Ann Oper Res 203:55–80

Dubey R, Gunasekaran A (2016) The sustainable humanitarian supply chain design: agility, adaptability and alignment. Intl J Logistics Res Apps 19:1–21

ElMaraghy HA, Mahmoudi N (2009) Concurrent design of product modules structure and global supply chain configurations. Int J Comput Integr Manuf 22:483–493

Farahani RZ, Rashidi Bajgan H, Fahimnia B, Kaviani M (2015) Location-inventory problem in supply chains: a modelling review. Int J Prod Res 53:3769–3788

Fazlollahtabar H, Mahdavi I, Mohajeri A (2013) Applying fuzzy mathematical programming approach to optimize a multiple supply network in uncertain condition with comparative analysis. Appl Soft Comput 13:550–562

Fleuren H, Goossens C, Hendriks M, Lombard MC, Meuffels I, Poppelaars J (2013) Supply chain--wide optimization at TNT Express. Interfaces 43(1):5–20

Ghayebloo S, Tarokh MJ, Venkatadri U, Diallo C (2015) Developing a bi-objective model of the closed-loop supply chain network with green supplier selection and disassembly of products: the impact of parts reliability and product greenness on the recovery network. J Manuf Syst 36:76–86

Goetschalckx M, Vidal CJ, Dogan K (2002) Modeling and design of global logistics systems: a review of integrated strategic and tactical models and design algorithms. Eur J Oper Res 143:1–18

Graves SC, Willems SP (2005) Optimizing the supply chain configuration for new products. Manag Sci 51(8):1165–1180

Habermann M, Blackhurst J, Metcalf AY (2015) Keep your friends close? Supply chain design and disruption risk. Decis Sci 46:491–526

Hearnshaw EJS, Wilson MMJ (2013) A complex network approach to supply chain network theory. Int J Oper Prod Manag 33:442–469

Hinojosa Y, Kalcsics J, Nickel S, Puerto J, Velten S (2008) Dynamic supply chain design with inventory. Comput Oper Res 35:373–391

Jamshidi R, Fatemi Ghomi SMT, Karimi B (2015) Flexible supply chain optimization with controllable lead time and shipping option. Appl Soft Comput 30:26–35

Kara SS, Onut S (2010) A two-stage stochastic and robust programming approach to strategic planning of a reverse supply network: the case of paper recycling. Expert Syst Appl 37:6129–6137

Kim Y, Choi TY, Yan T, Dooley K (2011) Structural investigation of supply networks: a social network analysis approach. J Oper Manag 29:194–211

Klibi W, Martel A (2012) Modeling approaches for the design of resilient supply networks under disruptions. Int J Prod Econ 135:882–898

Lambiase A, Mastrocinque E, Miranda S, Lambiase A (2013) Strategic planning and design of supply chains: a literature review. Int J Eng Bus Manag 5:1–11

Latha Shankar B, Basavarajappa S, Chen JCH, Kadadevaramath RS (2013) Location and allocation decisions for multi-echelon supply chain network - A multi-objective evolutionary approach. Expert Syst Appl 40:551–562

Li H, Womer K (2008) Modeling the supply chain configuration problem with resource constraints. Int J Proj Manag 26:646–654

Liu S, Papageorgiou LG (2013) Multiobjective optimisation of production, distribution and capacity planning of global supply chains in the process industry. Omega 41:369–382

Longo F (2012) Sustainable supply chain design: an application example in a local business retail. Simulation

Lorentz H, Kittipanyangam P, Singh Srai J (2013) Emerging market characteristics and supply network adjustments in internationalising food supply chains. Int J Prod Econ 145:220–232

Macchion L, Moretto A, Caniato F, Caridi M, Danese P, Vinelli A (2015) Production and supply network strategies within the fashion industry. Int J Prod Econ 163:173–188

Madadi A, Kurz ME, Taaffe KM, Sharp JL, Mason SJ (2014) Supply network design: risk-averse or risk-neutral? Comput Ind Eng 78:55–65

Maheut J, Besga JM, Uribetxebarria J, Garcia-Sabater JP (2014) A decision support system for modelling and implementing the supply network configuration and operations scheduling problem in the machine tool industry. Prod Plann Contr 25:679–697

Mansoornejad B, Pistikopoulos EN, Stuart PR (2013) Scenario-based strategic supply chain design and analysis for the forest biorefinery using an operational supply chain model. Int J Prod Econ 144:618–634

Medini K, Rabénasolo B (2014) Analysis of the performance of supply chains configurations using multi-agent systems. Intl J Logistics Res Apps 17:441–458

Meisel F, Bierwirth C (2014) The design of make-to-order supply networks under uncertainties using simulation and optimisation. Int J Prod Res 52:6590–6607

Meixell MJ, Gargeya VB (2005) Global supply chain design: a literature review and critique. Transport Res E Logist Transport Rev 41:531–550

Melo MT, Nickel S, Saldanha-da-Gama F (2009) Facility location and supply chain management - A review. Eur J Oper Res 196:401–412

Metta H, Badurdeen F (2013) Integrating sustainable product and supply chain design: modeling issues and challenges. IEEE Trans Eng Manag 60:438–446

Metters R, Walton S (2007) Strategic supply chain choices for multi-channel Internet retailers. Serv Bus 1:317–331

Mittermayer H, Rodríguez-Monroy C (2013) Evaluating alternative industrial network organizations and information systems. Ind Manag Data Syst 113:77–95

Mizgier KJ, Wagner SM, Jüttner MP (2015) Disentangling diversification in supply chain networks. Int J Prod Econ 162:115–124

Mohammadi Bidhandi H, Yusuff RM, Megat Ahmad MMH, Abu Bakar MR (2009) Development of a new approach for deterministic supply chain network design. Eur J Oper Res 198:121–128

Moncayo-Martínez LA, Zhang DZ (2011) Multi-objective ant colony optimisation: a metaheuristic approach to supply chain design. Int J Prod Econ 131:407–420

Nasiri GR, Davoudpour H, Karimi B (2010) A lagrangian-based solution algorithm for strategic supply chain distribution design in uncertain environment. Int J Inform Tech Decis Making 09:393–418

Nepal B, Monplaisir L, Famuyiwa O (2011) Matching product architecture with supply chain design. Eur J Oper Res 216:312–325

Nickel S, Saldanha-da-Gama F, Ziegler H-P (2012) A multi-stage stochastic supply network design problem with financial decisions and risk management. Omega 40:511–524

Olivares-Benitez E, Ríos-Mercado RZ, González-Velarde JL (2013) A metaheuristic algorithm to solve the selection of transportation channels in supply chain design. Int J Prod Econ 145:161–172

Osman H, Demirli K (2010) A bilinear goal programming model and a modified Benders decomposition algorithm for supply chain reconfiguration and supplier selection. Int J Prod Econ 124:97–105

Ouhimmou M, D'Amours S, Beauregard R, Ait-Kadi D, Chauhan SS (2009) Optimization helps Shermag gain competitive edge. Interfaces 39(4):329–345

Owen SH, Daskin MS (1998) Strategic facility location: a review. Eur J Oper Res 111:423–447

Paksoy T, Özceylan E (2013) Environmentally conscious optimization of supply chain networks. J Oper Res Soc 65:855–872

Pan F, Nagi R (2013) Multi-echelon supply chain network design in agile manufacturing. Omega 41:969–983

Parmigiani A, Klassen RD, Russo MV (2011) Efficiency meets accountability: performance implications of supply chain configuration, control, and capabilities. J Oper Manag 29:212–223

Pashaei S, Olhager J (2015) Product architecture and supply chain design: a systematic review and research agenda. Supply Chain Manag Int J 20:98–112

Paydar MM, Saidi-Mehrabad M (2015) Revised multi-choice goal programming for integrated supply chain design and dynamic virtual cell formation with fuzzy parameters. Int J Comput Integr Manuf 28:251–265

Petridis K (2015) Optimal design of multi-echelon supply chain networks under normally distributed demand. Ann Oper Res 227:63–91

Pirard F, Iassinovski S, Riane F (2011) A simulation based approach for supply network control. Int J Prod Res 49:7205–7226

Pokharel S (2008) A two objective model for decision making in a supply chain. Int J Prod Econ 111:378–388

Prakash A, Chan FTS, Liao H, Deshmukh SG (2012) Network optimization in supply chain: a KBGA approach. Decis Support Syst 52:528–538

Qu T, Huang GQ, Cung V-D, Mangione F (2010a) Optimal configuration of assembly supply chains using analytical target cascading. Int J Prod Res 48:6883–6907

Qu T, Huang GQ, Zhang Y, Dai QY (2010b) A generic analytical target cascading optimization system for decentralized supply chain configuration over supply chain grid. Int J Prod Econ 127:262–277

Qu T, Nie DX, Chen X, Chan XD, Dai QY, Huang GQ (2015) Optimal configuration of cluster supply chains with augmented Lagrange coordination. Comput Ind Eng 84:43–55

Quariguasi Frota Neto J, Bloemhof-Ruwaard JM, Van Nunen JAEE, Van Heck E (2008) Designing and evaluating sustainable logistics networks. Int J Prod Econ 111:195–208

Ramezani M, Bashiri M, Tavakkoli-Moghaddam R (2013) A new multi-objective stochastic model for a forward/reverse logistic network design with responsiveness and quality level. Appl Math Model 37:328–344

Ramezani M, Kimiagari AM, Karimi B (2014) Closed-loop supply chain network design: a financial approach. Appl Math Model 38:4099–4119

Ratha PC (2014) Towards concurrent product and supply chain designing: a review of concepts and practices. Int J Procure Manag 7:391

Rigot-Muller P, Lalwani C, Mangan J, Gregory O, Gibbs D (2013) Optimising end-to-end maritime supply chains: a carbon footprint perspective. Int J Logist Manag 24:407–425

Roni MS, Eksioglu SD, Searcy E, Jha K (2014) A supply chain network design model for biomass co-firing in coal-fired power plants. Transp Res Part E Logist Transp Rev 61:115–134

Santibanez-Gonzalez EDR, Diabat A (2013) Solving a reverse supply chain design problem by improved Benders decomposition schemes. Comput Ind Eng 66:889–898

Scott J, Ho W, Dey PK, Talluri S (2015) A decision support system for supplier selection and order allocation in stochastic, multi-stakeholder and multi-criteria environments. Int J Prod Econ 166:226–237

Selim H, Ozkarahan I (2008) A supply chain distribution network design model: an interactive fuzzy goal programming-based solution approach. Int J Adv Manuf Technol 36:401–418

Shabani N, Sowlati T (2013) A mixed integer non-linear programming model for tactical value chain optimization of a wood biomass power plant. Appl Energy 104:353–361

Shahzad KM, Hadj-Hamou K (2013) Integrated supply chain and product family architecture under highly customized demand. J Intell Manuf 24:1005–1018

Shi Y, Gregory M (1998) International manufacturing networks - to develop global competitive capabilities. J Oper Manag 16(2–3):195–214

Soleimani H, Seyyed-Esfahani M, Shirazi MA (2013) Designing and planning a multi-echelon multi-period multi-product closed-loop supply chain utilizing genetic algorithm. Int J Adv Manuf Technol 68:917–931

Sousa R, Shah N, Papageorgiou LG (2008) Supply chain design and multilevel planning-an industrial case. Comput Chem Eng 32:2643–2663

Srai JS, Gregory M (2008) A supply network configuration perspective on international supply chain development. Int J Oper Prod Manag 28(5):386–411

Su JCP, Chu C-H, Wang Y-T (2012) A decision support system to estimate the carbon emission and cost of product designs. Int J Precis Eng Manuf 13:1037–1045

Tian J, Yue J (2014) Bounds of relative regret limit in p -robust supply chain network design. Prod Oper Manag 23:1811–1831

Tiacci L, Paltriccia C, Saetta S, Martín García E (2014) Advances in production management systems. Innovative and knowledge-based production management in a global–local world. IFIP Adv Inform Comm Tech 440:563–570

Tiwari A, Chang PC, Tiwari MK (2012) A highly optimised tolerance-based approach for multi-stage, multi-product supply chain network design. Int J Prod Res 50:5430–5444

Ülkü S, Schmidt GM (2011) Matching product architecture and supply chain configuration. Prod Oper Manag 20(1):16–31

Vahdani B, Naderi-Beni M (2014) A mathematical programming model for recycling network design under uncertainty: an interval-stochastic robust optimization model. Int J Adv Manuf Technol 73:1057–1071

Vanteddu G, Chinnam RB, Gushikin O (2011) Supply chain focus dependent supplier selection problem. Int J Prod Econ 129:204–216

Verdouw CN, Beulens AJM, Trienekens JH, van der Vorst JGAJ (2011) A framework for modelling business processes in demand-driven supply chains. Prod Plann Contr 22:365

von Massow M, Canbolat M (2014) A strategic decision framework for a value added supply chain. Int J Prod Res 52:1940–1955

Wang F, Lai X, Shi N (2011) A multi-objective optimization for green supply chain network design. Decis Support Syst 51:262–269

Wilhelm W, Han X, Lee C (2013) Computational comparison of two formulations for dynamic supply chain reconfiguration with capacity expansion and contraction. Comput Oper Res 40:2340–2356

Wu C, Barnes D (2011) A literature review of decision-making models and approaches for partner selection in agile supply chains. J Purch Supply Manag 17(4):256–274

Wu T, Zhang K (2014) A computational study for common network design in multi-commodity supply chains. Comput Oper Res 44:206–213

Yadav SR, Muddada RRMR, Tiwari MK, Shankar R (2009) An algorithm portfolio based solution methodology to solve a supply chain optimization problem. Expert Syst Appl 36:8407–8420

Yang D, Jiao JR, Ji Y, Du G, Helo P, Valente A (2015) Joint optimization for coordinated configuration of product families and supply chains by a leader-follower Stackelberg game. Eur J Oper Res 246:263–280

Yao Z, Lee LH, Jaruphongsa W, Tan V, Hui CF (2010) Multi-source facility location-allocation and inventory problem. Eur J Oper Res 207:750–762

You F, Grossmann IE (2008) Mixed-integer nonlinear programming models and algorithms for large-scale supply chain design with stochastic inventory management. Ind Eng Chem Res 47:7802–7817

Zhang X, Huang GQ, Rungtusanatham MJ (2008) Simultaneous configuration of platform products and manufacturing supply chains. Int J Prod Res 46:6137–6162

Zhang Y, Bhattacharyya S, Li X (2010) From choice of procurement strategy to supply network configuration: an evolutionary approach. Int J Inf Technol Decis Mak 9:145–173

Zhang A, Luo H, Huang GQ (2013) A bi-objective model for supply chain design of dispersed manufacturing in China. Int J Prod Econ 146:48–58

Zokaee S, Jabbarzadeh A, Fahimnia B, Sadjadi SJ (2014) Robust supply chain network design: an optimization model with real world application. Ann Oper Res (online)

Chapter 4
Reconfigurable Supply Chains: An Integrated Framework

4.1 Introduction

The importance of supply chain management is increasing because companies face the necessity to improve customer service, which is not possible by considering just separate organizations. This need has been driven by increasing customer expectations, growing global competition, and technological developments, which have jointly contributed to greater uncertainty and volatility of enterprise management processes. However, as described in previous chapters, supply chain configuration, which forms the backbone of supply chain management, remains a long-term decision limiting the supply chain's ability to react to changing customer demand and operating environments.

The possibility to alter supply chain configuration with relatively minor resource requirements would be a desirable supply chain characteristic. It depends upon multiple factors, both logical and technological. However, intelligent decision-making and the ability to adequately implement decisions forms the basis for resolving problems associated with other factors. This chapter presents the concept of reconfigurable supply chains and outlines a general approach to enabling reconfigurability. The described approach puts forward the decision-making aspect and proposes model-integration as a cornerstone of efficient decision-making.

Section 4.2 introduces the concept of reconfigurable supply chains. It is followed by a description of multiple perspectives on supply chain configuration decision-making, presented in the form of supply chain configuration problem taxonomy. Finally, an integrated framework supporting supply chain reconfigurability is presented.

© Springer Science+Business Media New York 2016
C. Chandra, J. Grabis, *Supply Chain Configuration*,
DOI 10.1007/978-1-4939-3557-4_4

4.2   Reconfigurable Supply Chain

Reconfigurable supply chains is the next step in the evolution of supply chain structures. Forces driving this evolution and conditions for attaining reconfigurability are discussed. Reconfigurability also brings certain advantages and disadvantages to supply chain management. These advantages and disadvantages are also discussed in the following subsections.

4.2.1   Need

Modern supply chains have enabled enterprises to improve their performance by coordinating activities among supply chain members. Supply chain configuration has been the backbone of this cooperation, defining members involved in the supply chain and the physical and logical links among them. Establishing supply chain configuration is a long-term decision with a planning horizon of 2–5 years. Such a long-term orientation has enabled supply chain partners to implement highly efficient models of collaboration covering the entire product life cycle, starting with product design and ending with reverse logistics operations. Effective information exchange, process integration, materials and product movement, and collaborative planning mechanisms can be established and fine-tuned during the lasting cooperation. However, this approach has encountered multiple challenges in the last decade. Customer demand uncertainty is one of the primary challenges. This uncertainty shows up in multiple ways, such as increasing customer expectations for price, quality, and delivery performance; demand for customized products, shortened product life cycle, and erratic demand behavior. These factors are supplemented by traditional uncertainty concerning demand volume.

The customer demand satisfaction challenge is tightly related to increasing global competition and technology development challenges because these drivers encourage customers to ask for more. Global competition offers an increased number of alternative providers of goods and services. Additionally, characteristics of these goods and services such as price and quality, exhibit high variety. The technology development challenge offers less time to get acquainted with new technologies. On the other hand, technology development increases flexibility of manufacturing and service operations and simplifies the technical aspects of supply chain integration, which is an important enabler of efficient supply chain collaboration. However, this also allows companies to leave their current supply chain partners more easily and pursue involvement in other, more lucrative supply chains.

As a result, supply chains can no longer be expected to preserve their structure over a long horizon because they risk losing their competitiveness or face internal collapse. The supply chain configuration must be able to respond to changing customer demands and operating environments. Reinforcement and modification of supply chain configuration is one of the solutions of meeting these requirements. Therefore, appropriate mechanisms for supporting reconfigurability should be embedded in supply chain configuration decisions.

4.2.2 Definition

Reconfigurability is a required property to ensure supply chain agility and to enable service-oriented supply chains. For the purposes of this book, a reconfigurable supply chain is defined as:

> The reconfigurable supply chain is a supply chain possessing flexibility of altering its configuration with relatively minor resource requirements and without losing its operational efficiency in response to changing customer demands and operating environment.

By definition, a configurable (hence also reconfigurable) system can be designed, modeled, and configured for specific applications, and upgraded and reconfigured rather than replaced. With a reconfigurable system, new products and processes can be introduced with considerably less expense and ramp-up time. A wider interpretation of the supply chain reconfiguration term is simply changing the current supply chain configuration.

One example of supply chain reconfiguration is an ever evolving distribution network of Amazon.[1] Initially the company designed its distribution network with emphasis on cost savings, while recently it has switched its focus on improving delivery responsiveness by making long-term investments in fulfillment centers in highly populated urban centers.[2] On the other hand, to preserve agility; Amazon also uses short term leases of existing facilities to serve emerging markets and less established sales locations. For instance, while long-term facilities are built for serving customers in Florida, short term leased facilities are used in Nashville and Spain. Many examples of reconfiguration also can be found in the fashion industry. The clothing retailer GAP made significant changes in its supply chain configuration.[3] Some of the changes are made at the strategic level while others take place on a continuous basis at the tactical and operational level. At the strategic level, GAP emphasizes supply chain tailoring according to product and brand segmentation as

---

[1] MWPVL (2015), Amazon global fulfillment center network, http://www.mwpvl.com/html/amazon_com.html.

[2] SCDigest (2012), Amazon to Add 18 New Distribution Centers Worldwide in 2012. as It Keeps Investing in Logistics, http://www.scdigest.com/ONTARGET/12-08-07-1.php.

[3] Barrie, L. (2013), Supply chain key to GAP's global growth plans, Just-Style, http://www.just-style.com/analysis/supply-chain-key-to-gaps-global-growth-plans_id117581.aspx.

well as the omnichannel approach and supplier certification. At the tactical and operational level, the company, similarly as other agile fashion companies (Sen 2008), continuously optimizes order allocation among the preapproved supplier basis. It selects the right supplier for the current order to respond rapidly to customer requirements.

Supply chain configurability can be measured in mathematical models by a share of fixed investments; simulation allows to measure operational efficiency, especially with regard to ramp-up time.

From the supply chain network perspective, there are five main reconfiguration patterns:

- Reconfiguration by expansion
- Reconfiguration by contraction
- Reconfiguration by changing units
- Reconfiguration by changing links
- Reconfiguration by changing attributes of the units and links

The expansion pattern represents supply chain enlargement by adding more units to the supply chain. The contraction pattern represents removal of supply chain units. The reconfiguration by changing units indicates that either existing units are replaced with new units or type of the units changes. The reconfiguration by changing links pattern represents that the type of the links is changed or the links are added/deleted between existing supply chain units. These structural patterns are supplemented by the network level changes of supply chain attributes.

Figure 4.1 depicts the structural reconfiguration patterns. The graphical notation is used to show supply chain units added or removed as the result of reconfiguration, as well as supply chain links added or removed. This graphical representation is useful to show a to-be supply chain network after the reconfiguration assuming that the supply chain network shown in Fig. 2.3 illustrated the as-is supply chain network.

Another view of supply chain reconfiguration is distinction between the core supply chain structure and a reconfigurable part. Fig. 4.2 gives an illustration of this view along with notation used for these purposes. It is argued that some of the supply chain units and links are fixed over the current planning horizon and they are not supposed to be changed according to the current supply chain configuration strategy. On the other hand, there are supply chain units and links, which can be changed within the scope of the current configuration effort. Some of the units and links can only be changed at the design time while others can be changed during the supply chain execution as well. The design time refers to supply chain modeling phase (see Chap. 5) when a new supply chain configuration is decided upon. For instance, a decision is made to open a new plant and this decision cannot be easily reversed during the considered planning horizon. Actual supply chain operations are performed during the execution time. If a unit or a link is marked for selection at the execution time, it implies that the selection decisions are made dynamically. For example, there are several preapproved suppliers and materials can be ordered from any of these suppliers as necessary in the current operational situation.

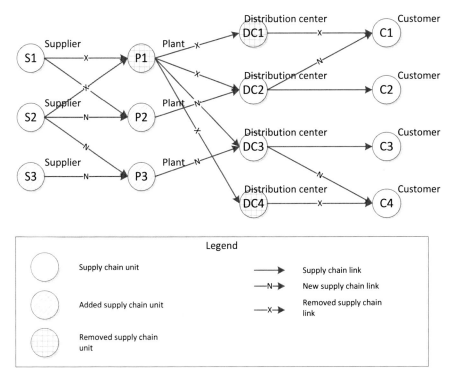

Fig. 4.1 Illustration of reconfiguration patterns

This representation defines options for supply chain reconfiguration on the basis of the core supply chain and is useful for decision-making purposes as later discussed in Chaps. 5 and 8.

4.2.3   Reconfiguration Example

The SCC Bike case is used to illustrate supply chain reconfiguration. The company considers three options for supply chain reconfiguration:

- Supplier NMPS2 could be treated as backup supplier to NMPS1 and this decision can be made during supply chain execution
- Supplier OIS1 could be used to supply add-on products for postponed customization directly to US DC distribution center
- Expansion to Canada is considered by either relocating US MW specialist retailer to service both regions or opening a dedicated specialist retailer CA ON in Canada.

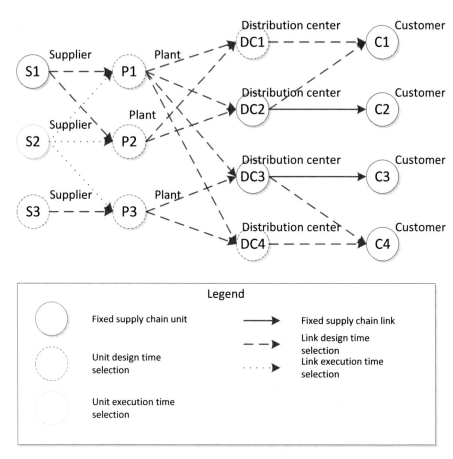

Fig. 4.2 Types of supply chain reconfiguration options

These reconfiguration options are shown in Fig. 4.3. Changes in other parts of the supply chain are not considered and corresponding units and links are shown as fixed. This reconfiguration example corresponds to the reconfiguration by expansion pattern.

Location of the specialist retailer serving Canadian market and allocation of this retailer are shown as design phase decisions. Several alternative supply chain configurations can be obtained by evaluation of the options. For instance, the expansion of Canada is deferred and the US MW retailer stays in its current location, Canadian and US MW customers are served by the relocated US MW specialist retailer US MW2 or US MW specialist retailer remains in its current location and a new CA ON specialist retailer is added for serving the Canadian market exclusively.

A particular configuration to be implemented is identified using supply chain configuration analysis methods as described in Chap. 5.

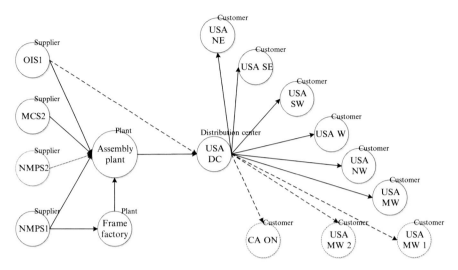

Fig. 4.3 Supply chain reconfiguration options for the SCC Bike example

4.2.4 Advantages and Difficulties

The problem of reconfigurable supply chains has three main aspects:

- *Decision-making*. Decisions about supply chain configuration are made, monitored and updated.
- *Physical implementation*. Building, opening, and operating manufacturing and service facilities, establishing and maintaining product flows, providing information technology infrastructure and designing and manufacturing reconfigurable products.
- *Logical implementation*. Business processes related to supply chain configuration and information systems support requirements are implemented.

The decision-making aspect determines what activities related to supply chain configuration are required. The physical infrastructure is built according to the decisions that are made. The logical implementation concerns utilization of physical implementation to achieve supply chain configuration and overall supply chain management objectives. It is also governed by the supply chain configuration decisions that are made.

The decision-making and logical implementation aspects are mainly afflicted by organizational difficulties and lack of knowledge. These deficiencies can be addressed by developing systematic and comprehensive decision-making and implementation procedures. Physical implementation is constrained by limited flexibility of available manufacturing technologies and high time and investment requirements. However, increasing use of outsourcing and third-party services in many situations eliminates the need for building an investment-heavy infrastructure. Similar improvements have also been achieved concerning manufacturing technologies.

The main advantages of reconfigurable supply chains are:

- *Robustness.* The supply chain is able to withstand external and internal shocks, such as loss of suppliers, labor disputes, and natural disasters, because suppliers can be replaced, manufacturing can be switched to alternative facilities, and transportation routes can be rearranged.
- *Flexibility.* Changing customer requirements can be accommodated by finding less expensive parts suppliers, choosing faster transportation channels, increasing product output volume, and introducing modified products.
- *Agility.* New business opportunities can be captured by engaging in relationships with innovative supply chain partners. Utilization of various Internet-based distribution options is a prominent example of supply chain redesign to find new business opportunities.

The main difficulties characteristic of reconfigurable supply chains and obstacles hampering their development are.

- *Organizational difficulties.* Time available to get accustomed to new partners, make decisions, and implement new business processes is limited. Lack of prior experience complicates decision-making and performance evaluation.
- *Technological constraints.* Manufacturing facilities may not support the processing of materials supplied by different suppliers and the production of different variations of products, or product design may not allow for easy modification and the relative independence of some of the parts suppliers.
- *Trust.* Partners may not engage in close collaboration and information sharing, partially because of the possibility that cooperation will be relatively short.

The reconfigurable supply chain assumes a dual position with regard to the lean manufacturing paradigm. It contradicts lean policies by maintaining extra capabilities needed to facilitate quick transitions from one configuration to another. For instance, flexible manufacturing equipment might be required despite lower efficiency compared to dedicated equipment. Possibility of frequent changes of configuration also hampers the fine-tuning of supply chain operations. On the other hand, reconfigurability requires keeping the supply chain simple and transparent, which coincides with requirements to achieve lean operations. For instance, many automotive companies are not able to restructure their manufacturing networks and increase efficiency of manufacturing operations because of highly entrenched labor agreements.

4.2.5   Requirements

Following the definition of the main supply chain configuration problem areas given above, the main requirements to be met to achieve reconfigurability are divided into two groups:

- *Technological requirements* covering the physical implementation aspect
- *Logical requirements* covering the decision-making and logical implementation aspects

The technological requirements concern aspect such as IT infrastructure, product design, and manufacturing and logistics technologies. The requirements on IT infrastructure imply that supply chain units should be able to exchange information and integrate processes. The product design requirements imply that product structure can be flexibly altered following changes in the supply chain configuration (i.e., replacement of parts suppliers). The manufacturing requirements imply that manufacturing technologies possess flexibility to change product mix and production volume. The logistics requirements imply that material and product distribution channels can be switched and that their capabilities are adjustable.

Significant progress has been made in meeting technological requirements for supporting reconfigurability. Requirements concerning IT infrastructure are discussed in Chap. 10 of this book. Product design and manufacturing and logistics technologies issues are discussed by Singhal and Singhal (2002), Koren et al. (1999), and Rehman and Subash Babu (2013). Modular product design allows replacing of components of products more easily. Therefore, suppliers can be substituted more easily, even though components that they supply are not physically identical to those used previously. Similarly, manufacturing automation systems allow for quicker adjustment to the new properties of materials used and products demanded by customers as well as reallocating manufacturing to other facilities. Finally, utilization of third-party logistics services allows for flexibility in choosing transportation channels, thus enabling cooperation with partners located across the globe and offering the required degree of delivery responsiveness.

Satisfaction of logical requirements is a challenging problem currently under active investigation. For purposes of further discussion, the following hierarchy of logical and business requirements is offered:

1. Commitment by entities involved in the supply chain.
2. Data and process integration.
3. Joint decision-making capabilities.
4. Joint decision-implementation and monitoring capabilities.
5. Data and process modification.
6. Modification of decision-making models.

Potential and existing supply chain partners must commit themselves to joint collaboration. Efficiency of decisions made often depends directly upon the willingness of supply chain members who need to agree on sharing potential supply chain benefits and losses, as well as sharing information and supporting cross-organizational business processes.

Data integration implies that consistent and current information necessary for decision-making and decision implementation is available within an organization, as well as across the supply chain. This requirement does not imply that all data need to be shared. Process integration implies that supply chain members are able to execute cross-organizational business processes. For instance, configuration decisions made by a system operated by one supply chain partner can be used to generate simulation-based decision evaluation models run by other supply chain partners or third-party logistics providers is automatically notified about replacement of a supplier to reroute shipments. Data and process integration enables joint decision-making, perceived as involvement of all supply chain members in the decision-making process which involves data gathering, decision-making, and analysis of results. Supply chain partners are informed about the judgment behind decisions made, which is important to provide some level of certainty to supply chain members engaged in a dynamic structure such as reconfigurable supply chains. There are two additional requirements concerning decision-making, such as representation of impact of uncertainty and treatment of temporal issues. These requirements imply that a reconfigurable supply chain is to be built with respect to stochastic influences and expected dynamic changes of the structure. Decisions need to be uniformly implemented across the supply chain. Data and process integration play important roles in achieving this requirement.

Finally, methods and tools for relatively inexpensive updating of data, processes, and models are needed as the supply chain is continuously reconfigured. Otherwise, the supply chain would lag behind planned changes.

The main attention in this book is devoted to business requirements, especially to joint decision-making capabilities.

4.3   Configuration Problems and Methods

The requirements defined above must be met for all supply chain management problems included in the supply chain configuration scope definition (see Chap. 2). The problem space is reduced by focusing mainly on the decision-making aspect of supply chain configuration, as defined in Sect. 4.2.

The supply chain management problems are classified as general and specific. General problems mainly deal with aspects of supply chain coordination and integration. Solving of a general problem includes solving several specific problems. Specific problems deal with a particular subject matter and can exist independently outside the supply chain environment. Supply chain configuration problems belong to the class of specific supply chain management problems. It involves multiple general and specific sub-problems. Furthermore, comprehensive evaluation of configuration decisions is not possible without considering interactions with other supply chain management problems.

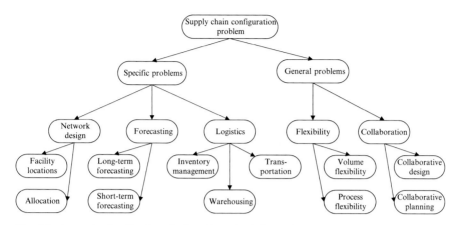

Fig. 4.4   A sample classification of selected supply chain configuration problems

To systemize accumulation and representation of supply chain management knowledge, a taxonomy of supply chain management and, particularly, supply chain configuration problems can be constructed. Development of a comprehensive, general taxonomy is a challenging task. However, supply chain members can develop their own taxonomy backed by industry-wide best practices as they accumulate supply chain management knowledge. Such taxonomy represents problems that a supply chain has dealt with and provides the basis for documenting problem-solving approaches. A segment of the supply chain configuration problem taxonomy is shown in Fig. 4.4. It shows only selected specific and general supply chain configuration problems. For instance, the network design problem includes aspects such as choice of network structure, selection of nodes, and establishing links among the nodes. If more detailed analysis is carried out, sub problems such as facility location and product-to-facility allocation are also addressed. All problems and sub problems can be further decomposed according to the circumstances characterizing these problems. For instance, the facility location problem is further specified using lower-level specific problems, such as static or dynamic facility location, or single or multiple facility location. Similarly, the short-term forecasting problem can be further decomposed according to criteria characterizing demand properties. This decomposition cannot be represented using simple linear classification trees. Classification tables categorizing low-level problems according to multiple criteria are needed.

Each problem can be addressed from multiple perspectives or views, such as data, process, space, and time. The data perspective characterizes the information required to make and implement supply chain configuration decisions. It also describes the structure of the supply chain configuration problem. The process perspective describes supply chain processes in relation to supply chain configuration. The space perspective addresses issues of locating supply chain units and other physical aspects of supply chain configuration. The time perspective allows analyzing of dynamic properties of supply chain configuration.

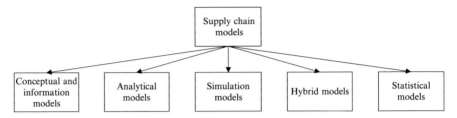

Fig. 4.5 Types of models used in supply chain configuration

Given the variety of problems and their evaluation perspective, no single model can cover all aspects of supply chain configuration. These methods can be represented using the taxonomy of supply chain configuration methods. This taxonomy is part of the overall supply chain management taxonomy and includes references to problem-solving methods interacting with configuration problem solving (Chandra and Tumanyan 2005).

Classification of supply chain models (see Fig. 4.5) constitutes the upper level of the taxonomy. Types of these models are:

- *Conceptual and information models.* These describe the supply chain configuration problem from a conceptual and information processing perspective. This category also includes IT-driven and business process reengineering models described in the literature.
- *Analytical models.* These mainly include mathematical programming models, which can be either deterministic or stochastic.
- *Simulation models.* These describe dynamic properties of supply chain configuration.
- *Hybrid models.* This combination of other types of supply chain configuration models are not necessarily confined just to combination of analytical and simulation models.
- *Statistical models.* Various statistical approaches are used to gain understanding about the supply chain configuration problem on the basis of accumulated historical data. These models so far are mainly considered as providing supporting functions, such as data preprocessing.

Further elaboration of taxonomy leads to identification of particular modeling methods for each type of models. However, this task is complicated by modeling methods belonging to various classes of models and methods. The taxonomy of supply chain methods would be a useful tool for identifying methods suitable for a particular decision-making situation. Some ideas for mapping between supply chain configuration problems and methods by using the supply chain taxonomy are presented in Chap. 6.

4.4 Integrated Framework

As indicated above, the book deals with decision-making aspects of the supply chain configuration problem. The identified requirements for successful supply chain configuration and reconfiguration in Sect. 4.2.5 state that model integration is one of the key aspects. Therefore, the supply chain configuration problem is addressed from the model integration perspective, with a general problem solving framework referred to as the Model Integration Framework (MIF). The framework naturally extends several related supply chain management and configuration frameworks.

4.4.1 Background

The need for integrating models representing various problems from different views is conceptually widely acknowledged. A large number of specialized integrated models have been developed. However, the integrated framework should support model integration in general. The integrated framework is developed in a spirit of a recent general drive for enterprise and extended enterprise integration, where decision-making is advanced as one of the main beneficiaries (Cummins 2002). This allows companies to achieve competitive advantage over other companies. Delen and Benjamin (2003) and Delen and Pratt (2006) actively promote a general integrated modeling framework that links enterprise description models, enterprise analysis models, and enterprise knowledge base. Von Massow and Canbolat (2014) position supply chain design in the overall corporate and supply chain strategy formulation and enactment framework. Evaluation of supply chain design decisions feeds back to updating the strategies.

Melnyk et al. (2014) analyze supply chain design along three dimensions, namely, influencers, design decisions, and building blocks. The framework focuses on identification of environmental factors affecting supply chain management. That can be perceived as definition of the supply chain management strategy and scope. Design decisions are constrained by the influencers and include physical and social network design as well as relationships governance and behavioral management. Finally, the building blocks such as inventory, transportation, capacity, and technology allow for implementing the decisions made.

Shapiro (2001) emphasizes that such a framework requires a tight integration between decision modeling and information technology support tools. The described supply chain optimization framework has a database management system as its central component. This system processes input data from corporate databases and maintains a supply chain decision database. The model generator is used to develop an optimization model using data provided by the database management system. The advanced optimizer is used to solve the optimization model.

Results obtained are stored in the decision-making database and are made available for further processing by spreadsheet programs and other analysis tools.

The supply chain configuration framework proposed by Dotoli et al. (2003) includes data analysis, network design, and solution evaluation modules. The data analysis module is used to preselect potential supply chain members by analyzing data accumulated in the company's database. The network design model is used to optimize supply chain structure. The solution evaluation module is used to evaluate the supply chain configuration by means of simulation. Evaluation is performed for various scenarios and informal feedback between evaluation and optimization is considered. Additionally, a search for consensus among decision-making parties at each decision-making stage is emphasized. Bounif and Bourahla (2013) describe an architecture of supply chain management decision support system, which combines optimization and simulation models and the basis of common supply chain model and decision-making rules. The system includes a performance evaluation module.

The decision-making and decision-implementation framework developed by Piramuthu (2005) specifically addresses the supply chain reconfiguration problem. Each supply chain unit dynamically chooses the available option for cooperation with supply chain partners. Decisions are based using the knowledge base in possession of each supply chain unit. Knowledge can be extracted from the knowledge base using various intelligent decision-making algorithms. The framework assumes that appropriate infrastructure is in place to implement any decisions made.

The object-oriented supply chain design modeling framework is developed by Kim and Rogers (2005). The modeling is driven by supply chain management goals and vision. The complex supply chain management problem is split into packages according to Supply Chain Operations Reference (SCOR) model division of supply chain process domains. Four views for each domain are developed to represent all aspects of the supply chain management problem. These views include function, structure (data), process, and behavior views, which are described using the Unified Modeling Language (UML) syntax. Business rules for transaction processing in the supply chain are added to the developed model. The obtained supply chain model can be used for implementation of a supply chain information system and are aimed to support relatively easy modification of this system. This framework emphasizes the information systems development aspect while the decision-making aspect is elaborated to a lesser degree.

4.4.2 Model Integration Framework

The key principle underlying the proposed supply chain configuration framework (see Fig. 4.6) is model synergy. The model synergy implies that each model complements others to provide different perspectives of supply chain configuration decision-making, and at the same time development and application of models is highly integrated to reduce complexity and to avoid inconsistencies and redundancies.

Fig. 4.6 The integrated
supply chain
reconfiguration framework

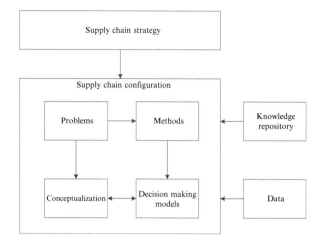

The proposed supply chain configuration framework starts with defining the decision-making capabilities of individual supply chain units on the basis of the common supply chain strategy and knowledge repository. The strategy outlines common goals of the supply chain. The repository defines common concepts pertinent to the supply chain configuration problem. The problem taxonomy and the methods taxonomy are developed using concepts defined in the repository. The knowledge model also contains the relevant supply chain standards, benchmarks, and other knowledge.

The supply chain configuration problems relevant to a particular decision-making situation can be mapped to problem-solving methods defined in the taxonomy of the supply chain configuration methods.

A supply chain configuration decision model can be developed by extracting appropriate decision-making methods from the taxonomy. To address the multi-dimensionality of the configuration problem, the decision modeling systems consist of multiple sub-models, which will be discussed in the following chapters. At the same time, conceptual and information models as a part of the supply chain problem conceptualization are used for a descriptive analysis of the decision-making problem, for defining the decision-making process, and data exchange mechanisms between the decision-making models and data sources. Similar to the decision-making models, conceptual models also represent different aspects of the supply chain configuration problem and are mutually interrelated. They use data extracted from problem and methods taxonomies to determine parts of the enterprise-wide information system that are relevant to a particular decision-making problem and the data that are needed to solve the problem.

The decision-making capabilities of individual supply chain members are brought together to enable joint decision-making and technological implementation of decisions (see Fig. 4.7). The supply chain configuration system brings together individual supply chain units. It is designed to support the proposed configuration framework (see Fig. 4.6). Each supply chain unit has its own supply chain

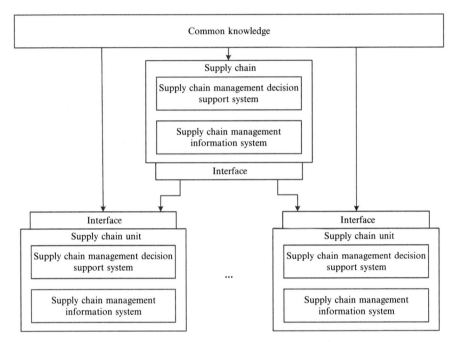

Fig. 4.7 Hierarchical relationships in supply chain reconfiguration

management system, which also implements the proposed framework as one of its modules. Supply chain configuration decisions are made through the collaboration of the overall supply chain configuration systems and supply chain management systems of individual units. The supply chain configuration system can be a centralized system maintained by one member or a group of supply chain members. In this case, a central supply chain configuration model is developed, which may also invoke models maintained by individual supply chain members. It can also be a distributed system, although such an approach appears to be more suitable for configuration monitoring and maintenance purposes.

The common knowledge refers to industry standards and generally accepted supply chain management concepts. The common knowledge facilitates established mappings between concept definitions in repositories maintained by individual supply chain units, thus leading to easier establishment of the common repository used by the supply chain configuration system.

An abstract interface is shown in Fig. 4.6. It provides data and process integration during both the decision-making process and the implementation of configuration decisions. Technological solutions for implementing this interface are discussed in Chap. 11.

Application of the proposed framework in the supply chain configuration process is described in the Chap. 5 which presents a supply chain configuration methodology, while practical implementations of the framework are discussed in Chaps. 7 and 10.

4.5 Summary

The proposed framework has brought together existing ideas on supply chain configuration decision-making and decision implementation. It also emphasizes concepts needed to support reconfigurability. The main features of the framework are:

- Modeling synergy and maintenance of consistent and up-to-date models
- Support for collaborative decision-making
- Utilization of decision-making capabilities of individual supply chain units as well as those of the entire supply chain
- Integration between decision modeling and the supply chain management information system
- Knowledge-driven approach
- Emphasis on efficient implementation of decisions

The framework enables reconfigurability by providing a means for efficient and comprehensive decision-making, streamlining the implementation of decisions and incorporating new members into the supply chain. This is mainly achieved by maintaining integrated and consistent information and decision-making models. Changes in the supply chain can be quickly represented into decision-making models and supply chain execution information systems.

Mapping between problems and available methods to a specific problem-solving model is performed by a decision analyst. Knowledge structuring and sharing is an important factor for facilitating this activity. Development of supply chain management ontology is an important step towards achieving interoperability (Grubic and Fan 2010). Cheung et al. (2012) show that the case based reasoning is a promising tool for supply chain integration.

It has been indicated that product design and manufacturing and logistics technologies play important roles in supporting reconfigurability. The literature review in Chap. 3 suggested that configuration decision-making models incorporating at least one of those aspects are currently of major interest. Wang et al. (2004) point out that supply chain design decisions should be driven by product characteristics and product life cycles. Blackhurst et al. (2005) propose a methodology for the design of supply chain operations by also considering the product and process design. Several other studies on coordinated product, process, and supply chain design are assembled by Rungtusanatham and Forza (2005) and Forza et al. (2005). An empirical investigation of manufacturing plans by Ortega Jimenez et al. (2015) shows that a combination of methods and technologies, such as JIT, TQ, HR, and TPM and improved organizational practices leads the transition from flexible manufacturing to reconfigurable systems.

References

Blackhurst J, Wu T, O'Grady P (2005) PCDM: a decision support modeling methodology for supply chain, product and process design decisions. J Oper Manag 23:325–343

Bounif ME, Bourahla M (2013) Decision support technique for supply chain management. J Comput Inf Technol 21:255–268

Chandra C, Tumanyan A (2005) Supply chain system taxonomy: a framework and methodology. Hum Syst Manag J 24:245–258

Cheung CF, Cheung CM, Kwok SK (2012) A knowledge-based customization system for supply chain integration. Expert Syst Appl 39:3906–3924

Cummins FA (2002) Enterprise integration: an architecture for enterprise application and systems integration. Wiley Computer Publishing, New York

Delen D, Benjamin PC (2003) Towards a truly integrated enterprise modeling and analysis environment. Comput Ind 51:257–268

Delen D, Pratt DB (2006) An integrated and intelligent DSS for manufacturing systems. Expert Syst Appl 30:325–336

Dotoli M, Fanti MP, Meloni C, Zhou MC (2003) A decision support system for the supply chain configuration. Proc. IEEE Int. Conf. Systems, Man and Cybernetics. Washington, DC. pp 1867–1872

Forza C, Salvador F, Rungtusanatham M (2005) Coordinating product design, process design, and supply chain design decisions: Part B. Coordinating approaches, tradeoffs, and future research directions. J Oper Manag 23:319–324

Grubic T, Fan IS (2010) Supply chain ontology: review, analysis and synthesis. Comput Ind 61:776–786

Kim J, Rogers KJ (2005) An object-oriented approach for building a flexible supply chain model. Int J Phys Distrib Logist Manag 35:481–502

Koren Y, Heisel U, Jovane F, Moriwaki T, Pritschow G, Ulsoy G, Van Brussel H (1999) Reconfigurable manufacturing systems. CIRP Ann Manuf Technol 48:527–540

Melnyk SA, Narasimhan R, DeCampos HA (2014) Supply chain design: issues, challenges, frameworks and solutions. Int J Prod Res 52:1887–1896

Ortega Jimenez CH, Machuca JAD, Garrido-Vega P, Filippini R (2015) The pursuit of responsiveness in production environments: from flexibility to reconfigurability. Int J Prod Econ 163:157–172

Piramuthu S (2005) Knowledge-based framework for automated dynamic supply chain configuration. Eur J Oper Res 165:219–230

Rehman AU, Subash Babu A (2013) Reconfigurations of manufacturing systems - an empirical study on concepts, research, and applications. Int J Adv Manuf Technol 66(1–4):107–124

Rungtusanatham M, Forza C (2005) Coordinating product design, process design, and supply chain design decisions: Part A: topic motivation, performance implications, and article review process. J Oper Manag 23:257–265

Sen A (2008) The US fashion industry: a supply chain review. Int J Prod Econ 114(2):571–593

Shapiro JF (2001) Modeling and IT perspectives on supply chain integration. Inf Syst Front 3:455–464

Singhal J, Singhal K (2002) Supply chains and compatibility among components in product design. J Oper Manag 20:289–302

von Massow M, Canbolat M (2014) A strategic decision framework for a value added supply chain. Int J Prod Res 52:1940–1955

Wang G, Dismukes JP, Huang SH (2004) Product-driven supply chain selection using integrated multi-criteria decision-making methodology. Int J Prod Econ 91:1–15

Chapter 5
Methodology for Supply Chain Configuration

5.1 Introduction

The previous two chapters highlight the magnitude of the supply chain configuration problem. Before starting with the description of models and tools available for solving the identified problems, a systematic approach for dealing with the configuration problem is laid out in this chapter. A systematic approach defined by a methodology would facilitate binding together different aspects of the configuration problem and provide problem-solving guidelines.

This chapter describes a general supply chain configuration methodology that aims to cover all major aspects of supply chain configuration. The methodology consists of eight steps. It starts with conceptual modeling of the supply chain configuration problems and gradually moves towards quantitative analysis. A description of the methodology includes guidelines and a set of methods and tools suited for performing specific steps. These methods and tools are outlined in this chapter and are discussed in more detail in Part II of the book. Given that solving the configuration problem requires support of multiple computational tools, architecture of the decision support system implementing the methodology is also developed.

The supply chain configuration methodology combines characteristics of methodologies used in operations research and business process management. The main processes comprising the methodology are identification and definition of the problem, solving the problem, implementation of decisions made, and continuous evaluation and monitoring. The configuration is performed as a continuous improvement cycle where changes are introduced in response to changes in the environment.

The rest of this chapter is organized as follows. Section 5.2 lays out the background for developing a methodology for supply chain configuration. Section 5.3 discusses the key issues to be addressed by the methodology. The entire configuration process is discussed in Sect. 5.4. Steps of the methodology are

© Springer Science+Business Media New York 2016
C. Chandra, J. Grabis, *Supply Chain Configuration*,
DOI 10.1007/978-1-4939-3557-4_5

elaborated in more detail in Sect. 5.5, which include outlining models available for performing each step. Architecture of the configuration decision support system is given in Sect. 5.6. Section 5.7 summarizes the chapter's contents.

5.2 Background

The methodology described in this chapter also draws upon several supply chain configuration methodologies proposed in the literature. Supply chain configuration methodologies typically cover decision-making stages, such as preparation for supply chain configuration problem solving, establishing the supply chain configuration, and an evaluation of decisions made.

The Cardiff Methodology for supply chain reengineering developed by the Logistics Systems Dynamics Group at the University of Wales (Naim 1996) is one of the first comprehensive methodologies for supply chain analysis at the strategic level. This methodology is primarily oriented toward an analysis of system dynamics using simulation as the basis of a supply chain business process model. It starts with defining business objectives followed by system input/output analysis. Construction of the conceptual model is the next step, aided by the library containing generic modeling components. The conceptual model is used in developing several quantitative models. Results obtained by means of quantitative modeling are verified and validated. Special attention is paid to model tuning and analysis of business scenarios. The methodology includes multiple feedback loops.

Ross et al. (1998) develop supply chain reconfiguration methodology, which focuses on the need for reconfiguration as a result of performance analysis of existing configurations. Best practices are identified during the first stage of the methodology. These are incorporated into the reconfigured supply chain. The authors list some of the methods available for performing each step. Consensus-building processes are emphasized.

Talluri and Baker (2002) have developed a three-phase supply chain configuration methodology. The first phase identifies and evaluates candidate supply chain units. The second phase establishes a supply chain configuration. The third phase deals with tactical planning on the basis of the established configuration. Mathematical models for each stage are provided. Two distinct features of this methodology are the presence of a broker representing the supply chain power structure and detailed discussion on the preselection of candidate supply chain units.

The supply chain configuration methodology developed by Dotoli et al. (2003) includes the creation of a decision-making team, data acquisition, preselection of candidate supply chain members using data envelopment analysis, and optimization and evaluation of the configuration. All major steps are followed by a discussion of the results. A multi-objective decision-making methodology developed by Chiang (2012) focuses specifically on design chain or network of partners joining efforts on product design. The methodology has three parts. The first part defines criteria for evaluation of design chain partners. The actual evaluation of the partners is performed in the second part. Integration of IT systems among the partners is one of the most important evaluation criteria. The performance evaluation of the design chain is performed in the third part of the methodology.

Piramuthu (2005) proposes a methodology for automated supply chain reconfiguration. This methodology is based on exploration of accumulated supply chain management knowledge. The knowledge based approach is also used by Prakash et al. (2012). The authors use the knowledge base to optimize supply chain network by means of genetic algorithms. Knowledge base initialization using data from various sources is a part of the approach.

Establishment of configuration and evaluation stages are considered in the methodology presented by Truong and Azadivar (2005). It is split into two parts: (1) determination of qualitative policy variables, and (2) determination of quantitative variables. These decisions are made in an iterative manner. Evaluation is performed using simulation modeling, where the simulation model is automatically generated according to optimization outcomes. A hierarchical top-down approach is taken by Corominas et al. (2015) in their SCOP methodology. It analyzes supply chain macrostructure, mesostructure and microstructure. The m-graph formalism is used for representation of the supply chain design.

The supply chain design problem in the changing environment through the prism of business-IT alignment is investigated by Medini and Bourey (2012). Their methodology uses SCOR as a reference for evaluation of the supply chain processes, and the five phase procedure for process optimization is elaborated.

General supply chain management modeling methodologies without particular focus on configuration are discussed by Simchi-Levi et al. (2007), Bowersox et al. (2012) and Chopra and Meindl (2012) among others. The supply chain planning methodology by Bowersox et al. (2012) consists of feasibility assessment, project planning, data collection, analysis (i.e., configuration modeling), development of recommendations, and implementation steps. Chopra and Meindl (2012) has supply chain network design as one of the steps of the overall supply chain management methodology.

These existing methodologies either consider the supply chain configuration problem as a part of the overall supply chain management process or focus on using a methodology specific network design model. The methodology elaborated in this section specifically deals with the supply chain configuration problem and is independent of particular network design models used.

5.3 Requirements

The methodology includes multiple steps common to many modeling methodologies, such as definition of performance measures, data gathering, execution of models, and analysis of results. There are also several issues relatively unique to supply chain environments and supply chain configuration. Many of these issues are defined by the distributed character of supply chains, which are often formed by relatively loosely coupled organizations.

The supply chain power structure is one of the major unique factors. The concept of a broker is adopted here (Ross et al. 1998). The supply chain configuration problem is formulated substantially differently depending on relationships between the organizations involved, thereby influencing parameters included in the model and considered performance measures.

The completely centralized supply chain owned by a single company is the most rigid case of the supply chain power structure. The decision-making process also can be centralized in the case of one dominating supply chain unit that picks its partners. However, even in this case, the dominant unit should account for the interests of other units to some extent because the notion of mutual dependence between supply chain members is widely recognized. In contrast to the centralized supply chain, a supply chain can be composed of independent units having approximately equal importance. In this case, the configuration decisions can be coupled with some compensation mechanisms, whereby some supply chain members compensate other members who bear additional configuration-related expenses to establish more efficient overall structure.

Although data gathering is the common function for any modeling effort, the distributed character of supply chains brings in an additional dimension to this problem. Data should be gathered not only from multiple sources in one organization, but also from multiple organizations what is made difficult by both technical and trust issues. Depending upon the configuration methods used data requirements vary greatly. However, even though mathematical programming seemingly requires only a little data, the data volume needed to estimate parameters accurately might be large. Data availability issues are softened by availability of integrated information systems, such as Enterprise Resource Planning (ERP) systems, and open Internet based data exchange standards.

Comprehensive evaluation of the configuration problem requires using several alternative models. It is well known that mathematical programming models are well suited to deal with the spatial aspects of a configuration problem while they struggle to deal with temporal aspects (Ballou 2001). Simulation is more appropriate to deal with the latter. The methodology should address the problem of efficient development of multiple models.

Appraisal of modeling results poses two major difficulties: (1) combination and interpretation of results given by multiple models; and (2) balancing quantitative results with assumptions made by a human decision maker. Multiple models present different views of the problem. It is crucial to evaluate results with respect to derived confidence bounds. Additionally, long-term strategic decisions involving huge costs and made by top executives are often adjusted on a judgmental basis. Although these adjustments representing factors not captured by models are often valuable, the balance between trust in quantitative results and judgmental decisions is to be defined to avoid nullifying the modeling effort.

To summarize this discussion, the main requirements for the methodology are outlined below:

- Parties involved and the power structure are clearly defined
- Data are well-structured to enable construction of multiple models
- Means for efficient selection of appropriate models and development of selected models are provided
- Guidelines for evaluation and approbation of modeling results are provided

The methodology is also required to provide guidelines for addressing organizational issues, and to support the development and maintenance of a modeling repository, which accumulates information about decision-making processes.

5.4 Configuration Steps

The supply chain configuration methodology forms a cycle of continuous improvement in response to changes in the environment and new business opportunities. It is applicable in the case of new supply chain development as well as in the case of reconfiguration of the existing supply chain. In the former case, a new cycle is started with the supply chain strategy updating step without feedback from existing supply chain operations. In the latter case, a new cycle is started if deviations from the specified supply chain configuration performance measurements are observed or new business opportunities are identified during the monitoring and evaluation step.

Given that the supply chain configuration initiative has been initiated, the supply chain configuration process follows the methodology outlined below:

Step 1—Supply chain strategy updating. The supply chain configuration problem is solved as part of the overall supply chain strategy, which is updated according to the changes in the environment. Relationships between the strategy and the supply chain configuration are identified and objectives and the scope of the supply chain configuration problem are defined at the executive level.

Step 2—Conceptual modeling. A formal definition of the supply chain configuration problem under consideration is established. It describes supply chain configuration objectives, concepts and processes as descriptive supply chain models. These models can be used in qualitative supply chain analysis and more importantly they are used as the unified basis for development of other supply chain configuration models. The conceptual model is also used to structure data required for the supply chain configuration problem solving, and the appropriate data are also gathered during this step.

Step 3—Experimental planning. In order to achieve supply chain configuration objective stated in Step 2, appropriate modeling and analysis methods are selected and experimental scenarios are defined. All parties involved also agree on acceptance criteria for modeling results and procedures for adjusting modeling results by human decision makers.

Step 4—Preselection. Supply chain configuration is typically established from a set of candidate units, possibly of different types. This step reduces the number of candidate units. It is necessary to reduce computational burden at the following selection step and because different selection criteria are often used at the preselection stage.

Step 5—Modeling and analysis. Interrelated qualitative and quantitative supply chain configuration models are developed and applied according to the experimental plan established in Step 3. Models are used to explore different aspects of the supply chain configuration problem. Additionally, combinations of models (i.e., hybrid models) are often considered. Verification and validation of models is also a part of this step.

Step 6—Decision-making. The Modeling and analysis yields multiple alternative supply chain configurations and their evaluation results under different conditions. Multi-criteria and group decision-making methods as well as strategic management methods are used to select the most appropriate supply chain configuration for implementation

Step 7—Implementation. Physical location of supply chain units and establishing of flows among units is the main concern of implementation. However, there are many logical aspects, as well. Adoption of configuration decisions triggers development of models for tactical and operational decision-making. Some of the configuration models or their parts might be used in a continuous manner.

Step 8—Monitoring and evaluation. Day-to-day transactions are executed in the supply chain after changes in the supply chain configuration have been implemented. The key performance indicators relevant to the supply chain configuration are monitored, and the alignment among the supply chain strategy, expected performance and capabilities supported by the supply chain configuration is continuously evaluated. Identification of misalignment leads to initiation of the new supply chain configuration cycle.

The supply chain configuration methodology is graphically illustrated in Fig. 5.1.

5.5 Elaboration of Steps

This section elaborates various steps in the methodology. For every step, its purpose, main tasks, methods, and outcomes are defined. The main properties of the methods are described below, while a detailed description of the supply chain configuration methods is given in Part II of this book.

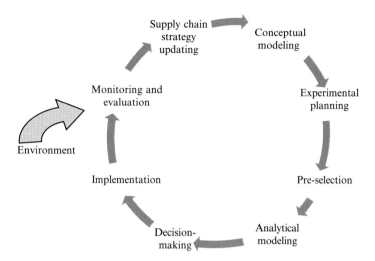

Fig. 5.1 Steps of the supply chain configuration methodology

5.5.1 Supply Chain Strategy Updating

The supply chain strategy updating step provides an initial description of the supply chain configuration problem. Outcomes of this step provide a basis for further formalization of the decision-making problem during the conceptual modeling step.

The purpose of this step is to define the goal of supply chain configuration in alignment with the overall supply chain strategy and to prepare and execute description of the supply chain configuration problem to be addressed. The main tasks of this step are: (1) updating of the supply chain strategy and identification of interaction between the updates and the supply chain configuration; (2) identification of the supply chain configuration decision-making circumstance; and (3) definition of the supply chain configuration scope.

Strategic supply chain management methods are used to define the supply chain strategy. The templates developed in Chap. 2 of this book are used to describe the decision-making circumstances and the supply chain configuration scope. The main outcome of this step is an agreement among all parties involved on objectives of the supply chain configuration effort.

5.5.1.1 Supply Chain Strategy

The supply chain strategy takes into account multiple factors such as customers, competition, corporate goals, global environment, technology development and risks to create new or enhance existing supply chain capabilities. It is also affected by the monitoring and evaluation results of the existing supply chain configuration.

Table 5.1 Key aspects of the SCC bike supply chain strategy

Aspect	Description
Customers	Cycling enthusiasts
Corporate goals	Proving high level customer support in a peer-to-peer fashion
	Increasing use of bicycles for recreation and transportation
Business strategy	Growth in new markets
Supply chain strategy	Flexible supply chain

The supply chain can pursue one of the two main business strategies, namely, the growth strategy and the efficiency improvement strategy. In doing so, it can adopt lean, flexible, agile, service-oriented, or mixed supply chain strategies. As an example, Table 5.1 lists key aspects of the supply chain strategy for the SCC Bike supply chain.

The supply chain strategy as a whole influences all supply chain management processes. Therefore, relationships between the strategy and the supply chain configuration problem solving are described to ensure alignment. The alignment evaluation is performed along with identification of decision-making circumstances and configuration scope definition.

5.5.1.2 Decision-Making Circumstance

Table 5.2 lists attributes characterizing decision-making circumstances. The supply chain configuration initiative is driven by changes in the supply chain strategy and drivers for change initiating supply chain reconfiguration. The drivers for change are case specific and can be categorized as those due to new product development, supply chain interrelationships, competitive pressures and environmental changes. The supply chain interrelationships characterize state of supply chain units and link among the units. For instance, a unit is taken over by a competitor or when location or transportation mode becomes unavailable.

Values of the power structure attribute are similar to those of the broker. However, the decision-making situation is defined by a combination of power structure and broker. For instance, the supply chain configuration initiative is put forward by a broker representing the minority unit in the supply chain environment with the dominating unit. The specialized dominating unit power structure implies that a dominating unit concentrates just on its core competencies while the nonspecialized dominating unit power structure implies that a dominating unit assumes various functions and different supply chain stages. For instance, many automotive Original Equipment Manufacturers (OEM's) are transforming themselves from a nonspecialized dominating unit to a specialized dominating unit by outsourcing the manufacturing of many components and abandoning plans to enter distribution.

Decision-making is greatly influenced by the initial state of the supply chain. The initial state influences collaborative decision-making processes because the

Table 5.2 Attributes characterizing supply chain configuration decision-making circumstances

Attribute	Values
Driver of change	New product development
	Supply chain interrelationships
	Competitive pressures
	Environmental changes
Power structure	Specialized dominating unit
	Nonspecialized dominating unit
	Supply chain wide consortium
	Consortium of several units
	Equal power units
Initial state of the network	New supply chain
	Existing supply chain with minority of units fixed
	Existing supply chain with majority of units fixed
Information sharing	Complete information sharing
	Limited information sharing
	No-information sharing
Data availability	Historical records available
	No historical records available
	Some historical records available
Number of alternatives	No alternatives
	Few candidates
	Large number of alternatives

level of trust among potential partners might vary substantially. This attribute also relates to the information availability attribute. In the case of a new supply chain, little information is available for appraisal of parameters characterizing links between supply chain units. For instance, a potential supply chain partner can evaluate its delivery lead time and quote it for supply chain modeling purposes. However, this quote may not account for specific time delays caused by interactions between this particular supplier and a manufacturer.

More efficient supply chain decisions can be made and implemented if information is shared among supply chain partners. In the case of complete information sharing, the main problem is establishing physical and logical channels of information exchange. The complete information sharing applies only to information needed for supply chain configuration decision-making and implementation of decisions. In the case of limited or no information sharing, a broker relies on publicly available data, indirect observations (e.g., historical sales orders), and assumptions. Use of indirect observations leads to problems such as the bullwhip effect (Lee and Whang 2000). Limited information sharing is frequently a problem in early decision-making stages, and one should assess whether information sharing will improve upon engagement in supply chain execution.

Data availability relates to information sharing. However, even if complete information sharing is in place, historical data might not be available. This problem is especially severe in the case of the design of a completely new supply chain network. Many statistical analysis methods used in the preselection stage depend upon data availability, which varies among supply chain partners. Therefore, accuracy of estimates for individual units needs to be taken into account during evaluation of the supply chain network.

The number of alternatives characterizes such factors as number of alternative suppliers, number of alternative locations for manufacturing and distribution facilities, and number of transportation modes. The number of alternatives substantially influences the selection of decision-making models. A large number of alternative suppliers usually require preselection of suppliers. A large number of alternative locations requires initial continuous search for optimal locations. Abundance of alternatives complicates data gathering and model-solving tasks. The product variety also needs to be accounted for. Aggregation of products is usually considered in the case of high product variety.

The power structure directly influences the definition of configuration objectives. For instance, a dominating unit can focus almost exclusively on its own objectives, while a consortium of equal partners needs to consider the objectives of all partners in the multi-objective framework. A choice of relevant parameters and costs is particularly influenced by a broker. For instance, from the point of view of a manufacturer, he/she is only concerned about the cost of raw materials, but not about the other internal costs of a potential supplier. Questions about accounting for transportation costs also need to be resolved—either these are included in the purchasing cost, paid by the supplier, or paid by the manufacturer.

Decision-making problem definition also depends upon the type of broker and the power structure. For instance, a broker representing a minority unit also analyzes costs incurred to the dominating unit, even if he/she does not account for these costs directly.

5.5.1.3 Scope

The definition of scopeprovides an initial description of the decision-making problem. This initial description will be formalized and further refined in the next step by means of conceptual modeling. Issues represented in the scope definition correspond to those identified in Chap. 2. The SCC Bike supply chain configuration scope definition is given in Table 5.3 as an example. The configuration efforts span all supply chain stages, though majority of decisions are to be taken at the supply and distribution tiers. The configuration is performed with respect to supply chain costs and customer service to be evaluated both at the strategic and tactical decision-making levels. The decisions and parameters considered in this supply chain configuration initiative are some of the most commonly used.

The scope definition is aligned to the flexibility strategy. Specifically the configuration effort concerns supplier selection to ensure reliable supplies and focuses on ensuring high level of customer service which is one of the key problems in bicycle supply chains.

Table 5.3 SCC bike supply chain configuration scope

Scope parameter	Values
Objectives and criteria	Cost minimization Customer service
Horizontal extent	All supply chain stages
Vertical extent	Strategic Tactical
Decisions	Supplier selection Supply chain planning DC location Customer allocation
Parameters	Demand Capacity Fixed costs BOM
Processes and functions	Source Make Deliver

5.5.2 Conceptual Modeling

The conceptual modeling is the first step towards formal modeling and analysis of the identified supply chain configuration problem. It elaborates a comprehensive descriptive representation of the supply chain configuration problem and provides the basis for further supply chain configuration activities.

The purpose of this step is to formalize definition of the supply chain configuration problem and gathering of appropriate data. The main tasks are: (1) definition of supply chain configuration goals, relevant concepts and processes; (2) data structuring; and (3) data gathering from heterogeneous data sources.

The enterprise modeling technique 4EM (Sandkuhl et al. 2014) is adopted for definition of goals, concepts, and processes. This technique allows to represent enterprises by a number of interlinked views each describing particular aspect of the enterprises. The 4EM technique provides both the model development language and enterprise modeling guidance. It promotes participative modeling practices and consensus seeking among all parties involved. Therefore, it is well-suited also for decentralized supply chain configuration problem solving. The interrelated enterprise models developed are stored in the modeling repository and can be retrieved for updating in the case of new supply chain reconfiguration cycle.

Chapter 7 describes application of the 4EM technique in supply chain management. The key feature of this application is that initially a generic supply chain model is built for the whole supply chain configuration domain. The generic model is reused and augmented during conceptual model building for a specific supply chain configuration case.

A data model is derived from the conceptual model to serve as the basis for data integration. The data model defines all objects, parameters, and costs relevant to the specific supply chain configuration problem. Values are assigned to these parameters and costs using mapping between data sources and supply chain objects instantiated from the defined classes. Types of data sources are:

- Actual data—e.g., transportation costs as quoted by providers
- Empirical estimates—e.g., processing times if no automated recording is available
- Theoretical estimates—e.g., capacity costs and inventory holding costs are often difficult to estimate. Theoretical and judgmental approaches can be used in this case
- Forecasts—e.g., demand forecasts
- Transactional data—e.g., processing time, if automated recording of operations is performed. Aggregates of transactional data are typically used

Memory depth (i.e., how many historical records are available) of transactional data is also important. For additional information on data processing issues, data types, and their sources; readers are referred to Shapiro (2006).

5.5.3   Experimental Planning

The conceptual modeling step produced the formal descriptive representation of the supply chain configuration problem. In order to proceed with further evaluation, an experimental plan is developed. The purpose of experimental planning is to define procedures for modeling and analysis of the supply chain configuration problem. The tasks of experimental planning are (1) selection of appropriate modeling methods; (2) definition of performance measures; (3) identification of relevant experimental scenarios and experimental factors; and (4) definition of individual experiments to be conducted as well as their properties.

Multiple modeling methods are used for modeling purposes. The common data model is used to enable usage of common information among the models. A decision-making workflow is developed to automate experimentation. The workflow defines a sequence of modeling activities as well as a way the models interact among themselves. Currently, the workflow definition is usually hardcoded, though recently there is a movement towards more flexible approaches to model integration (Levis 2015).

Completion of these tasks depends upon the type of models used for selection purposes, such as statistical, knowledge-based, optimization, and simulation (see Fig. 5.2). The types of models are discussed in detail in Part II of this book.

Statistical models are mainly used for preselection purposes. However, they can be used for establishing the final supply chain configuration as well, if further refinement of results is not deemed possible or necessary. Availability of information is crucial to applicability of statistical models. In Fig. 5.2, Statistical models are designated for situations with a large number of alternatives and a few fixed units. Knowledge-based and optimization models still should perform better in this

Fig. 5.2 Selection of configuration methods according to the modeling scope

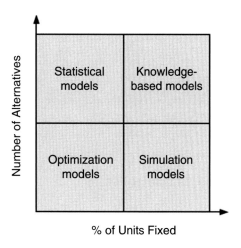

quadrant, although model development and solving, respectively, could be a major obstacle for applicability of these models.

Knowledge-based models (Piramuthu 2005; Choy et al. 2002) are well-suited in situations where the base supply chain has been established and inclusion of a limited number of units needs to be regularly reassessed.

However, optimization and simulation models remain key model types used in the selection step of supply chain configuration. Optimization models are used to establish the supply chain configuration if the number of alternatives is large. Simulation can be used if the number of alternatives is small. It is also preferable that some of the units are fixed to allow for easier validation of simulation models. Additionally, both models are frequently used together. Optimization models are used to establish the supply chain configuration by evaluating a large number of alternatives, while simulation is used to evaluate optimization results in a more detailed manner according to multiple evaluation criteria (for more detailed discussion on differences between optimization and simulation, see Part II of this book and Law and Kelton (2014)). A combination of optimization and simulation models does not exclude other model combinations. Hybrid modeling, where optimization and simulation models are tightly integrated, is often used to attain benefits from using both optimization and simulation.

There is a large variety of optimization models used in supply chain configuration. A majority of them are mixed-integer programming models. Stochastic and multi-objective programming models are also becoming popular. Similarly, discrete-event simulation is the dominant simulation modeling technique used for supply chain configuration.

The development of simulation models typically is more time consuming than the development of optimization models. The same applies to application of models. However, for many large-scale problems, direct solving of optimization models can also be time consuming. Therefore, specialized solution procedures need to be developed for solving optimization models. The proposed methodology and supply chain configuration framework, emphasizes automated model building. That is especially important if simulation modeling is used to evaluate optimization results. To avoid repeated development of a simulation model for each optimization outcome, simulation models can be automatically generated according to optimization results.

The outcome of the step is the experimental plan to be executed using the selected modeling methods.

5.5.4   Preselection

The conceptual models prescribe the types of units needed, their functions, and evaluation criteria. During this step, a list of potential candidates for each type of units is compiled and candidates are evaluated according to the evaluation criteria.

The purpose of the preselection step is to narrow the set of alternative configurations. The tasks of this step are (1) identification of potential supply chain units; and (2) preselection of the most promising supply chain units. A data driven multi-criteria evaluation method is used for the preselection.

Data availability for conducting evaluations is a major concern during the preselection step. Candidate units have different data-sharing policies and it is also expensive for a broker to collect extensive data for a large number of candidates. The level of data availability may vary for each criterion, causing difficulties in assessing the accuracy of a composite criterion. The broker must prioritize data requirements and pay attention to units responsible for most important processes and to key criteria. Table 5.4 provides classification of some of the data availability situations. A majority of models reported in literature pay little attention to information availability and sharing issues during the preselection and selection steps. Generally, it is assumed that all necessary information (e.g., material prices, lead times) is available. For instance, Li and O'Brien (1999) build their supply chain design model under an assumption that partners provide all necessary information to evaluate their performance. Chapter 11 on data driven supply chain configuration discusses modern approaches to data integration for preselection purposes.

Table 5.4 Classification of preselection situations

	New supply chain	Existing supply chain
Information sharing	Partners' performance data are available but benchmarking basis might not be available	Performance requirements can be identified from current performance data and compared to those supplied by partners
Limited information sharing	Preselection is based upon assumptions and indirect observations	Indirect observations can be evaluated with regard to actual performance observations

A number of methods are available for preselection. These include ranking and weighting, benchmarking, statistical analysis, data envelopment analysis, analytical hierarchy process, and several artificial intelligence-based methods.

The preselection and qualification phase is reviewed by Wu and Barnes (2011) who identify various decision-making models used during this stage. Weber et al. (1991) and De Boer et al. (2001) reference articles using linear weighting and statistical methods for the evaluation of suppliers. Analytical Hierarchy Process (AHP) is used for supplier preselection according to multiple criteria, although a similar procedure could be applied for preselection of other supply chain stages, too (Vaidya and Kumar 2006). This method is promoted as a more systematic approach compared to simpler weighting methods. It describes decomposing a complex problem into a multi-level hierarchical structure of objectives, criteria, and alternatives. AHP starts with the identification of criteria influencing decision-making. A hierarchy of the criteria is built. Each criterion is compared with all other criteria according to a specified scale to assert its relative importance. Consistency of the assessment is checked. As a result, a relative importance weight is obtained for each criterion and can be applied to rank suppliers.

Wang et al. (2004) use AHP to obtain an aggregated AHP weight for each candidate supplier according to multiple criteria, which have varying degrees of importance. The key first-level criteria are delivery reliability, flexibility and responsiveness, cost, and assets. The weights computed are used afterwards for final supplier selection. Fuzzy hierarchical process is used to account for some of the uncertainty in the evaluation process (Chiang 2012).

Choy et al. (2002) use case-based reasoning and neural networks to evaluate and benchmark potential suppliers. Performance of these evaluation methods depends upon data provided by potential suppliers and availability of historical data. The disadvantage of artificial intelligence-based methods is their lack of generality, and subsequently only basic features are usually used.

Data envelopment analysis (Charnes et al 1994) is used to evaluate the efficiency of supply chain units by Ross et al. (1998) and Talluri and Baker (2002). Data envelopment analysis (DEA) is a method for determining and benchmarking the efficiency of decision units such as potential supply chain partners. Each decision unit has a number of outputs converted into output performance indicators. Efficiency is measured as a ratio between the weighted sum of outputs and the weighted sum of inputs. The efficiency for each unit is optimized relative to other units by finding optimal values of the weights for the given unit. The efficiency measure can be used to make preselection decisions concerning the given unit. In both papers referred to above, efficiency measures for candidate units are computed. These measures are afterwards used in optimization of the supply chain configuration. Talluri and Baker (2002) additionally split between determining the number of units needed and actual product quantities assigned to each unit.

The special case of preselection is the continuous facility location method. If facilities can be located in a large number of locations, continuous facility location methods (Drezner and Hamacher 2002) identify appropriate location area, and a specific location can be chosen among several alternative locations in this area.

5.5.5   Modeling and Analysis

Supply chain configuration modeling is performed using qualitative and quantitative models according to the design of experiments. The purpose of this step is to obtain a number of good supply chain configurations for the final decision-making in the next step of the methodology. The tasks of the modeling and analysis step include (1) development of models; (2) experimentation; (3) verification and validation of the models; and (4) analysis of results. The methods used in these tasks depend upon their selection during the experimental design. These include process analysis heuristics, mathematical programming, simulation, and hybrid methods. These are presented in more detail in Part II of this book.

Validation of configuration models is difficult even in the case of existing supply chain reconfiguration, because the feedback loop between decision-making and implementation results is long and often obscure. Therefore, expert judgment is one of the main approaches to results validation. Additionally, application of multiple models can also be used as a form of validation. The situation is better if a majority of units are fixed and the effect of replacing individual units can be easily observed through performance measures, such as on-time delivery and material cost.

The outcome of this step is a number of alternative supply chain configurations characterized by their performance measurements and the results of sensitivity analysis.

5.5.6   Decision-making

The modeling and analysis step yields multiple alternative supply chain configurations. The purpose of this step is to select the most appropriate configuration for implementation. The decision-making step includes definition of evaluation criteria, comparison of the alternatives according to these criteria and what-if analysis.

In majority of cases, decision-making is performed by human decision-makers, though in agile and service-oriented supply chain, increasingly the configuration decisions are made routinely in an automated manner. In order to facilitate the decision-making process, means for visualization and what-if analysis are provided. Visualization is provided using a geographical information system, which shows the alternative supply chain configurations in the wider context.

5.5.7   Implementation

Acceptance of results, pertains to a managerial decision-making problem area (Maccrimmon and Taylor 1976). One of the key problems during this step is interpretation of results given by multiple models and distilling the final decision and contingency plans. Malhotra et al. (1999) discuss theoretical aspects of decision-making using multiple models. A framework for decision-making using multiple models is provided. It shows that the final decision is a combination of outputs from various models adjusted by managers' judgmental decisions. Ouhimmou et al. (2009) specifically identifies roles and stakeholders involved in configuration decision-making processes from the practical experience. They include vice president of manufacturing, production, procurement, sales, and logistics managers. Methods for consensus building are also incorporated in some decision support systems (Tung and Turban 1998; Limayem and DeSanctis 1999)

Implementation of configuration decisions includes dealing with both logical and physical aspects of implementation, as discussed in Chap. 4. Physical implementation is beyond the scope of this book, although this issue has not received adequate coverage in literature. Information technology-related issues of implementation are discussed in Chap. 11.

5.5.8 *Monitoring and Evaluation*

The performance of the established configuration needs to be monitored and evaluated according to decision-making criteria used during decision-making, and key performance indicators, which are important to the individual units and the entire supply chain. Traditional engineering control mechanisms, such as control charts, can be used (Montgomery (1996) for description of monitoring methods). Monitoring of appropriate performance measures can be implemented using a link between supply chain transactions processing, and decision-making provided by information modeling data-mapping functionality. Abu-Suleiman et al. (2004) describe a framework for supply chain performance management. The framework is based on the Balanced Scorecard approach, and uses performance metrics defined in the SCOR model.

The monitoring results are evaluated and decisions concerning necessary changes in the supply chain configuration are made. These decisions are based on an assessment of the current supply chain situation (Bowersox et al. 2012). The situation assessment includes internal review of supply chain structure and performance, assessment of market and competition, evaluation of the relationship between supply chain partners, and assessment of technological factors. This situation assessment is further refined during the initialization step.

In Part III of this book, where applied studies are discussed, approbation and the impact of configuration decision-making results are discussed in more detail.

5.6 Architecture of Decision Support System

The computational complexity of the configuration problem requires the assistance of a comprehensive decision support system. The decision support system provides not only core services like data integration, model building, and model solving, but

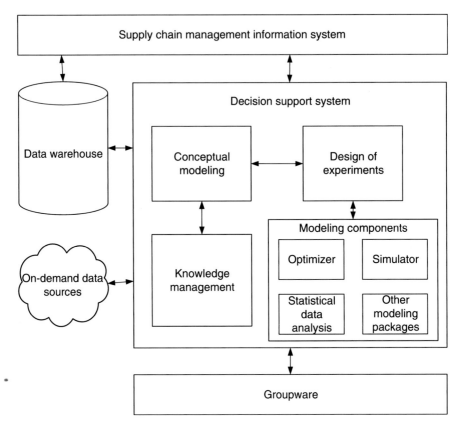

Fig. 5.3 Components of the supply chain configuration decision support system

also supporting services for organization of the decision-making process, maintaining modeling repository, and preliminary analysis of results. Additionally, the decision support system must be incorporated in the overall supply chain management system.

Fig. 5.3 shows a schematic representation of the architecture of the decision support system. This architecture complies with the integrated framework described in Chap. 4 and provides a complete set of tools for carrying out the supply chain configuration methodology.

The decision system consists of components for exploring supply chain configuration from various perspectives. The four key components are Knowledge management, Conceptual modeling, Design of experiments and Decision-modeling. The decision-modeling component is used to make configuration decisions mainly associated with Steps 5 and 6 of the methodology. The conceptual modeling component is responsible for maintaining different views of descriptive representation of the supply chain configuration problem (Step 3). Another major function of this component is linking the decision-modeling system with the supply chain management information system. The knowledge management module maintains generalized information for conceptualizing the specific supply chain configuration problem. The experiment planning module manages the configuration problem exploration by defining scenarios to be explored and gathering modeling results for final decision-making.

The key components are supported by data processing and storage components and groupware. The decision-support system is also a part of the overall supply chain management system and is integrated with the supply chain information system. The supply chain information system is a system tracking all supply chain transactions and containing references to other decision-making applications. It is one of the major data sources for decision-making purposes. Additionally, it supplies the conceptual modeling component with existing enterprise and supply chain information models and meta-data. Data sources are shown separately because besides information from the supply chain management information system, external data sources are also used. Data warehousing is used for organizing and presenting data from various sources for decision-making purposes (Kimball et al. 2008). However, not all data need to be stored in data warehouses because some of them have small volume and are rather specific. Usage of on-demand data sources is becoming more prevalent and data are often retrieved and only temporary stored for exploring specific decision-making scenarios. The groupware module is included to orchestrate the decision-making process and to support collaborative decision-making.

Sample implementations of this architecture are discussed in Chap. 10 of this book, along with a general discussion of IT tools that support supply chain configuration decision-making and decision implementation.

5.7 Summary

The supply chain configuration methodology has been presented in this chapter. It is aimed at addressing a wide range of issues arising during decision-making and the implementation of decisions. The methodology emphasizes the importance of model integration to enable the supply chain configuration problem for a comprehensive evaluation. Additionally, efficiency of model development is also stressed. However, the extent to which the methodology is applied varies from case to case (i.e., not all steps are always required, and the importance of each step also varies).

Another aspect of the methodology is that it aims to position the supply chain configuration problem solving in the overall supply chain decision-making and decision-implementation framework, as well as in relation to the overall supply chain management information system. Information modeling is the major mechanism in achieving this integration. If integration with other decision-making processes and the supply chain management system is not an important objective, then information modeling can be accomplished in a less formal manner.

Application of the methodology is not possible without using various decision-modeling tools. The architecture of the supply chain configuration decision support system described in this chapter depicts the main tools used and the relationships among these tools. Support for collaborative decision-making and the accumulation of knowledge are emphasized in the architecture. Again, the supply chain configuration decision support system should share many components with other modeling applications and rely on efficient use of the supply chain information technology infrastructure.

References

Abu-Suleiman A, Boardman B, Priest JW (2004) A framework for an integrated supply chain performance management system. IIE Annual Conference and Exhibition, pp 613–618

Ballou RH (2001) Unresolved issues in supply chain network design. Inf Syst Front 3:417–426

Bowersox DJ, Closs DJ, Cooper MB (2012) Supply chain logistics management. McGraw-Hill, Boston

Charnes A, Cooper WW, Lewin AJ, Seiford LM (1994) Data envelopment analysis: theory, methodology and applications. Kluwer, Boston

Chiang TA (2012) Multi-objective decision-making methodology to create an optimal design chain partner combination. Comput Ind Eng 63:875–889

Chopra S, Meindl P (2012) Supply chain management: strategy, planning, and operation. Prentice Hall, Upper Saddle River

Choy KL, Lee WB, Lo V (2002) An intelligent supplier management tool for benchmarking suppliers in outsource manufacturing. Expert Syst Appl 22:213–224

Corominas A, Mateo M, Ribas I, Rubio S (2015) Methodological elements of supply chain design. Int J Prod Res 1–14

De Boer L, Labro E, Morlacchi P (2001) A review of methods supporting supplier selection. Eur J Purch Supply Manag 7:75–89

Dotoli M, Fanti MP, Meloni C, Zhou MC (2003) A decision support system for the supply chain configuration. In: Proc. IEEE Int. Conf. Systems, Man and Cybernetics, Washington, DC, pp 1867–1872

Drezner Z, Hamacher H (2002) Facility location: applications and theory. Springer, Berlin

Kimball R, Ross M, Thornthwaite W, Mundy J, Becker B (2008) The data warehouse lifecycle toolkit. John Wiley, New York

Law AM, Kelton WD (2014) Simulation modeling and analysis. McGraw-Hill, New York

Lee HL, Whang S (2000) Information sharing in supply chain. Int J Technol Manag 20:373

Levis AH (2015) Multi-formalism modeling of human organization. Proceedings 29th European Conference on Modelling and Simulation, pp 19–31

Li D, O'Brien C (1999) Integrated decision modelling of supply chain efficiency. Int J Prod Econ 59:147–157

Limayem M, DeSanctis G (1999) Providing decisional guidance for multicriteria decision making in groups. Inf Syst Res 11:386–401

Maccrimmon KR, Taylor RN (1976) Decision making and problem solving. In: Dunnette M (ed) Handbook of industrial and organizational psychology. Rand-McNally, Chicago, pp 1397–1453

Malhotra MK, Sharma S, Nair SS (1999) Decision making using multiple models. Eur J Oper Res 114:1–14

Medini K, Bourey JP (2012) SCOR-based enterprise architecture methodology. Int J Comput Integr Manuf 25:594–607

Montgomery DC (1996) Introduction to statistical quality control. Wiley, New York

Ouhimmou M, D'Amours S, Beauregard R, Ait-Kadi D, Chauhan SS (2009) Optimization helps Shermag gain competitive edge. Interfaces 39(4):329–345

Naim MM (1996) Methodology before technology. Manuf Eng 75(3):122–125

Piramuthu S (2005) Knowledge-based framework for automated dynamic supply chain configuration. Eur J Oper Res 165:219–230

Prakash A, Chan FTS, Liao H, Deshmukh SG (2012) Network optimization in supply chain: a KBGA approach. Decis Support Syst 52:528–538

Ross A, Venkataramanan MA, Ernstberger KW (1998) Reconfiguring the supply network using current performance data. Decis Sci 29:707–728

Sandkuhl K, Stirna J, Persson A, Wißotzki M (2014) Enterprise modeling: tackling business challenges with the 4EM method. Springer, Berlin

Shapiro JF (2006) Modeling the supply chain. Duxbury Press, New York

Simchi-Levi D, Kaminsky P, Simchi-Levi E (2007) Designing and managing the supply chain. McGraw-Hill/Irwin, New York

Talluri S, Baker RC (2002) A multi-phase mathematical programming approach for effective supply chain design. Eur J Oper Res 141:544–558

Truong TH, Azadivar F (2005) Optimal design methodologies for configuration of supply chains. Int J Prod Res 43:2217–2236

Tung L-L, Turban E (1998) A proposed research framework for distributed group support systems. Decis Support Syst 23:175–188

Vaidya OS, Kumar S (2006) Analytic hierarchy process: an overview of applications. Eur J Oper Res 169:1–29

Wang G, Huang SH, Dismukes JP (2004) Product-driven supply chain selection using integrated multi-criteria decision-making methodology. Int J Prod Econ 91:1–15

Weber CA, Current JR, Benton WC (1991) Vendor selection criteria and methods. Eur J Oper Res 50:2–18

Wu C, Barnes D (2011) A literature review of decision-making models and approaches for partner selection in agile supply chains. J Purch Supply Manag 17(4):256–274

Part II
Solutions

Chapter 6
Knowledge Management as the Basis of Crosscutting Problem-Solving Approaches

6.1 Introduction

In Chap. 2, we argue that supply chain configuration is one of the principal supply chain management decisions and that it has a profound impact on other subsequent managerial decisions. As described therein, the supply chain configuration problem is a complex problem, which is composed of several sub-problems. It is also emphasized that the solutions to these problems require design, modeling, and problem-solving techniques based on knowledge from various fields such as systems science, systems engineering, operations research, industrial engineering, decision sciences, management science, statistics, information sciences, computer science, and artificial intelligence. Some of the prominent techniques utilized from these fields are information modeling, process modeling, simulation modeling, data mining, and optimization. We build on this proposition by adopting a key problem of information integration in the supply chain, which has an embedded structure representing various sub-problems, and how its management relates many of the concepts espoused in this book about supply chain configuration. Also, this problem serves as a prime example of how crosscutting approaches drawn from various disciplines highlighted above may be adopted in devising solutions for the complex supply chain configuration problem. Before we proceed further, let us first develop a clear understanding of the information integration problem in the supply chain.

The Supply Chain and the Information Integration Problem. One way to look at a supply chain is as an alignment of firms that bring products or services to the market (Lambert et al. 1998). This alignment is in the form of an extended enterprise, where firms collectively organize the supply, production, and distribution of products and services.

The management of such a complex organization can be brought to the integration of its business processes. Process-oriented management vs. function-oriented management is an important feature that makes the supply chain a distinct enterprise system class. Another facet of supply chain system complexity is its

© Springer Science+Business Media New York 2016

C. Chandra, J. Grabis, *Supply Chain Configuration*,

DOI 10.1007/978-1-4939-3557-4_6

organizational dynamics and operational specifics. Organizational dynamics assumes frequent changes in organizational structures such as control hierarchy, goal structure, members' network, and so on. Operational specifics are mainly related to the uncertainty in which a supply chain organization operates. Integration of supply chain processes assumes additional complexity when the decision-making mode (i.e., centralized vs. decentralized) is considered in the mix.

One of the key issues in managing a supply chain process is information integration among its constituents. To facilitate this integration, supply chain information resources ought to be effectively organized and shared. Information integration provides channels that convey information from one supply chain constituent to another. One form of this problem involves the integration of existing implementations that have been built in heterogeneous infrastructures, such as different hardware platforms, operating systems, and database management systems. Presenting the data on which applications perform in a uniform, self-consistent way ensures that they share the same view of the supply chain. Another form of integration is concerned with working collectively on common problems by sharing an understanding of the problems' reasoning logic and applying best practices. This provides a common architecture in information sharing so that supply chain members' collaborative activities provide performance improvement to each member and to the entire supply chain.

The problem of process integration, and its surrogate information integration described above, needs an appropriate solution. In this context, we advocate the necessity of applying system principles and knowledge management methodologies based on the following reasons: (1) the extent of knowledge becomes intractably large, (2) business units are geographically decentralized but more closely networked, (3) collaboration among individual workers is important, and (4) challenges are faced in eliciting requirements when user partners are large, decentralized, and unknown.

In this chapter, we propose a framework and implementation mechanisms for designing a knowledge management system capable of supporting organizational dynamics and operational uncertainty, as well as facilitating process integration in a supply chain. Taxonomies and ontologies are viewed as a means for conceptualizing the knowledge to share and utilize in decision-modeling applications. They bring formalism into the knowledge management system, thus offering standards for communication, which is necessary for collaborative problem solving.

The general trend in process integration is to develop information models that system users can share, thereby sharing the same view of the world (CIMOSA (Kosanke 1995), TOVE (Gruninger et al. 2000), Supply chain ontology (Grubic et al. 2011). The gap seen in these and other research efforts is the absence of a system reference model that can identify information model components and define mechanisms for their design and implementation. This reference model formally represents the source system, such as supply chain, its informational needs, and constructs that need to be built to support system processes. The ultimate target and value of the proposed approach, and hence the reference model is taxonomy and ontology development as a platform for integrated supply chain knowledge management. That is particularly important in the era of web-based supply chain collaboration and cloud computing, where the reference models and integration standards provide the basis of knowledge sharing and supply chain integration (Huang and Lin 2010).

Further, the proposed reference model enables fulfilling the purpose of this chapter in laying the ground work for integrated solutions proposed in Chaps. 7–9. The importance of this chapter is to highlight that solutions to supply chain configuration problems must integrate complex modeling and analysis techniques drawn from a host of disciplines.

The chapter is organized as follows. Section 6.2 describes the motivation, focus, and significance of crosscutting approaches. Section 6.3 discusses the notion of taxonomy and ontology and how it contributes to system integration. Section 6.4 introduces the knowledge management system development framework, which starts from the source system, goes through system component decomposition, and presents knowledge-modules design for these components. Section 6.5 formally presents the knowledge management system reference model, relating elements in the proposed framework and describing their meaning. Section 6.6 presents four stages of the knowledge management system development life cycle, as well as describes how the reference model can be implemented for each stage.

6.2 Crosscutting Approaches: Motivation, Focus, and Significance

Supply chain configuration draws from an array of fields as far as framework, models, and methodologies are concerned. This is primarily owing to the impact of any configuration effected on a supply chain, on its strategic, tactical, and operational decision-making environments. In this section, we discuss the motivation behind developing an integrated supply chain configuration framework and a reference model for designing knowledge and its management, with the aim of improving supply chain management.

6.2.1 Motivation and Focus

The motivation and focus of the research methodology proposed in this chapter is to integrate various problem-solving approaches from a host of fields in the design of proposed supply chain configuration problem-solving methodologies. It is characterized by two main purposes: general and specific.

The *general* purpose is to develop a common body of interdisciplinary knowledge to understand issues and problems related to reconfigurable systems.

The *specific* purpose is to (a) develop methodology and tools for supply chain reconfiguration, (b) elaborate framework for knowledge-based problem analysis and model building, and (c) quantify factors influencing supply chain reconfiguration.

The general problems in reconfigurable systems can be classified as related to the system's environment, availability of appropriate modeling tools, interconnectedness of decisions at various levels of supply chain, and availability of common knowledge throughout the system. These can be listed as follows:

- Increasing competitive pressures and consumer focus requires innovative supply chain modeling and management tools.
- Supply chain modeling tools must capture complex interactions within the supply chain.
- Supply chain configuration decisions have significant impact on other decisions at all levels.
- Knowledge assumes a critical role in a firm's success, and, therefore must be captured, organized and utilized effectively.

Problem-solving strategies applied to reconfigurable manufacturing systems entail developing (a) domain independent solution(s) templates at the macro level, (b) capability models for application specific domain dependent problems at the micro level, and (c) coordination models to integrate models developed in (a) and (b).

6.2.2 Problem Solving for Configurable Systems

To provide an integrated overview of interconnectedness of crosscutting research areas for configurable systems, three problem-solving approaches are proposed: systemic, reductionist, and analytic. These are defined as follows:

- *Systemic Approach* This incorporates the abstract level. This *level of inquiry* deals with issues of scalability of system, meta-modeling of systems, and defining the dynamic knowledge problem domain model.
- *Reductionist Approach*. This incorporates the activity level. This *level of inquiry* consists of dynamic knowledge problem domain model, internal state, and goals and objectives of the enterprise *units,* (producer, plant, department, supplier, vendor, etc.), and strategic management models.
- *Analytic Approach*. This incorporates the implementation level. This *level of inquiry* consists of internal state, goals and objectives of the enterprise units, strategic management models, and shared goals and objectives of the enterprise.

We discuss below related research in direct comparison to these three problem-solving approaches.

At the systemic level, a supply chain is a general class of system that exhibits a cooperative behavior within its business and market environment (Klir 1991).

The foundation of this system is built on a network architecture that has various demand and supply nodes as it provides, as well as receives goods and services to and from its customers and suppliers, respectively (Chandra 1997; Lee and Billington 1993; Swaminathan et al. 1998; Bellamy and Basole 2013). Supply chain system frameworks describe general foundational elements of integration between its marketing and production functions. These are in the form of general theories, hypotheses, standards, procedures, and models that are based on well-founded principles in these disciplines (Cohen and Lee 1989; Deleersnyder et al. 1992; Drew 1975; Graves 1982; Hackman and Leachman 1989; Lee 1993; NIST 1999; Tzafestas and Kapsiotis 1994; Younis and Mahmoud 1986). Systems modeling deals with general modeling issues of this class of systems, such as how to represent, quantify, and measure cooperation, coordination, synchronization, and integration (Little 1992; Morris 1967; Pritsker 1997). Systems engineering describes methodologies for structuring systems as these are implemented in various application domains (Blanchard and Fabrycky 1990). System integration deals with achieving common interface within and between different components at various levels of hierarchy in an enterprise (Shaw et al. 1992), as well as different architectures and methodologies (ISO TC 184/SC 5/WG 1 1997; IMTR 1999; Hirsch 1995), using distributed artificial intelligence and intelligent agents (Gruber 1995; Stumptner 1997; Wooldridge and Jennings 1995).

At the reductionist level, a supply chain configuration must be based on its local, as well as global, environmental constraints. These constraints are partly imposed as the supply chain negotiates and compromises to adapt to its cooperative behavior (Jennings 1994). Enterprise modeling as a technique has been used effectively in decomposing complex enterprises, such as a supply chain. Ontologies are defined to describe unique system descriptions of supply chains that are relevant to specific application domains (Gruninger 1997). The classic problem for a supply chain is an inventory management problem requiring coordination of product and information flows through a multi-echelon supply chain. This class of problem has been solved by integration of the front and back ends of the supply chain with costs and lead times as key measures of its performance (Clark 1972; Clark and Scarf 1960; Diks et al. 1996; Diks and De Kok 1998; Hariharan and Zipkin 1995; Pyke and Cohen 1990).

The analytic approach for the general class of supply chains has its origins in economic models of supply and demand coordination. Game Theory principles for payoffs among market competitors have been used effectively to design competitive strategies for supply chains (Gupta and Loulou 1998; Masahiko 1984). Coordination and cooperation—dealing with interfaces between strategies, objectives, and policies for various functions of an enterprise, has received much attention in optimizing the performance of a supply chain (Malone and Crowston 1994; Thomas and Griffin 1996; Whang 1995). Various aspects of cooperation have been prescribed for effective management of supply chains (Sousa et al. 1999).

Starting from the evaluation of existing enterprise integration architectures (CIMOSA, GRAI/GIM, and PERA), the IFAC/IFIP Task Force on Architectures for Enterprise Integration has developed an overall definition of a generalized architecture framework called GERAM or Generalized Enterprise Reference Architecture and Methodology (ISO TC 184/SC 5/WG 1 1997).

6.2.3   Significance of This Approach

Supply chain management strategies have the potential of enabling smart manufacturing and services

- Adaptable, and integrated equipment, processes, and systems that can be readily reconfigured.
- Manufacturing processes that minimize waste.
- System synthesis, modeling, and simulation for all manufacturing operations.
- Technologies to convert information into knowledge for effective decision-making.
- Software for intelligent collaboration systems.
- New educational and training methods that enable the rapid assimilation of knowledge.

The common thread in the deployment of these technologies is achieving (a) reconfigurability, (b) efficiency, and (c) complex modeling and analysis in decision-making related to managing advanced manufacturing systems.

This emphasis on developing enhanced manufacturing capabilities and technologies to support infrastructure mandates research in following crosscutting areas

- Adaptable and reconfigurable manufacturing systems.
- Information and communication technologies.
- Processes for capturing and using knowledge for manufacturing.
- Adopting and incorporating IT into collaboration systems and models focused on improving methods for people to make decisions, individually and as a group.
- Enterprise modeling and simulation.
- Analytical tools for modeling and assessment.
- Managing and using information to make intelligent decisions among a vast array of alternatives.
- Adapting and reconfiguring manufacturing enterprises to enable formation of complex alliances with other organizations.

The objective of the research presented in this chapter is to formalize the capture and management of supply chain management knowledge accumulated in various domains of science, engineering, and technology, and using various problem-solving techniques.

6.3 Taxonomy, Ontology, and System Integration

To see linkages between problems and decision-making models utilized in a complex enterprise such as a supply chain, it is imperative that these components be formally represented. Taxonomy and ontology provide the means to classify the supply chain problems and represent formal knowledge, which is used in decision-making. We take up discussion on this topic next.

6.3.1 Taxonomy

According to the American Heritage® Dictionary of the English Language, Fourth Ed. (2000), taxonomy is the classification of organisms in an ordered system that indicates natural relationships. It is the science, laws, or principles of classification. Further, it is an arrangement by which systems may be divided into ordered groups or categories according to common characteristics.

System taxonomy reflects information about relationships both inside the system, and with its surrounding environment. Supply chain system taxonomy aims to provide a multidisciplinary representation of supply chain activities and characteristics. The review of research in the field of supply chain taxonomy development reveals that most of them are based on single case studies, providing taxonomy for a subset of information. System taxonomy is organized for the entire system. Organizing information representation for a part of a system or for one problem jeopardizes decision-making because it may miss some key aspects. The supply chain is an organization whose components are interrelated to each other. This cohesion makes the system unmanageable if it is considered as one unbreakable unit. Based on biological classification, system taxonomy provides mechanisms for dividing a supply chain system into relatively independent units, providing as minimal a coupling between units as possible by collecting characteristics in groupings by their similarity. Further, iterative decomposition of groupings and creating new groupings can build a robust hierarchy of describing system characteristics.

System taxonomy serves two purposes: (1) standardization of terms and definition, and (2) unification of information representation. This brings out reusability of developed information models, as well as organization and structure, to knowledge management. Scalability and traceability are the most important features that system taxonomy provides, and thus, new features can be added and existing ones easily found.

6.3.2   Ontology

In the Artificial Intelligence (AI) literature, ontology is defined as the study of the kinds of things that exist (Sowa 2000). In AI, programs and logic deal with various kinds of objects, and we study what these kinds are and their basic properties (McCarthy 2003). Over the years, ontology has become more than an abstract representation of objects and their properties and is becoming a part of the software application domain with application to other branches of AI, such as heuristics and epistemology. The latter is a study of the kinds of knowledge that are required for solving problems in the world, and the former is a way of trying to discover something, or an idea embedded in a program. Along with shaping its pragmatic purpose, ontology has found its application in many fields, such as knowledge representation, system integration, enterprise modeling, conceptual modeling, and Semantic Web.

The above definition of ontology by Sowa (2000), as a study of the kinds of things that exist, is very generic. However, during the last two decades, several features of ontology have evolved that define its broader and more diverse scope and purpose in designing information support for decision-making. A review of the pertinent literature offers the following contrasting definitions and interpretations of ontology to validate our above assertions:

- Ontology is an explicit specification of conceptualization (Gruber 1993), meaning that ontology defines kinds of things, their possible relationships, and plausible implementation.
- Ontology is a catalog of types of things that are assumed to exist in a domain of interest, D, from the perspective of a person who uses a language, L, for the purpose of talking about D (Sowa 2000). This feature of ontology assumes the existence of a language with enough expressiveness for representing the domain of interest.
- Ontology refers to an engineering artifact, constituted by a specific vocabulary that is used to describe a certain reality and by a set of explicit assumptions regarding the intended meaning of words in the vocabulary (Guarino 1995). This definition of ontology adds a new feature requiring that it must have mechanisms and terminology for describing the meaning of words and vocabulary as well as their interpretations.

Ontology as a tool for information modeling has been adopted for a large body of research initiatives. As part of the research described in this chapter, a number of ontology tools, languages, and research projects have been studied to understand the role of ontology in information support systems, particularly for information integration. Some of them, such as Ontolingua (Farquhar et al. 1997) and OntoBroker (Fensel et al. 2001), investigate ontology narrowly as a standalone discipline. Other projects, such as TOVE (Fox and Gruninger 1999; Fox et al. 2000) and DOGMA (Meersman 2001), combine knowledge organization with specific domains, investigating agents for which knowledge is organized. Others look at the problem more widely, including development of ontologies in enterprise modeling systems, such as Enterprise (Stader 1996) and Process Handbook (Malone et al. 1999), aiming to support organizations effectively in change management. What is common in all of these projects, however, is that ontology explicitly defines the vocabulary presented with a language in which queries and assertions are exchanged among users (Grubic and Fan 2010). The next section describes the knowledge management system development stages in a framework.

6.4   Knowledge Management System Development: A Proposed Framework

The framework for knowledge management system conceptualization is depicted in Fig. 6.1. Based on the theoretical background developed in Sect. 6.3, a technique is proposed for conceptualizing supply chain organization and problem knowledge. The proposed advances offer integration of knowledge components with decision support systems and their consumption by software applications or agents. Knowledge components encompass ontology models and the infrastructure supporting their creation, storage, and use.

To serve the needs of a knowledge management system in formalizing and delivering knowledge to decision-modeling applications, several requirements are imposed on ontology conceptualization, such as (1) systematic principles for knowledge conceptualization, (2) the problem-specific nature of ontology constructs, (3) the modularity and object nature of formed knowledge, (4) reusability of created knowledge, (5) integration of distributed data, and (6) machine-readable format of delivered knowledge.

The genesis of the proposed framework is taxonomy and its amplifications to problems and problem-solving techniques, particularly when applied to supply chain management.

6.4.1   Taxonomy Development

Taxonomy is a systematic representation of a system's existence (McKelvey 1982). Accordingly, taxonomy is built based on principles of system theory. It is a mechanism for structuring the knowledge about a certain system domain. The process of taxonomy development consists of information collection, systematic analysis, and classification of system attributes.

Problem taxonomy provides the overall framework under which problem-oriented information system components can be designed and implemented.

Supply chain problem taxonomy comprises: (a) classification of supply chain problems, (b) classification of problem solving methodologies for supply chain

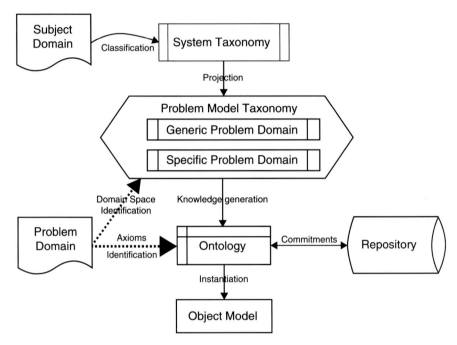

Fig. 6.1 The conceptual framework of a knowledge management system

management, and (c) hierarchical classification of variables or factors necessary for dealing with such problems. We explain this concept with the help of one of the fundamental problems in the supply chain management literature—the bullwhip effect.

The Bullwhip Effect
The most downstream supply chain unit observes an external demand, transmitted up on a supply chain as inventory replenishment orders move from one unit to another. It has been observed that substantial information distortion may occur during this transmission. This information distortion, known as the bullwhip effect, appears as an order variance increase as one moves up the supply chain.

Classification of the Bullwhip Effect Problem
The bullwhip effect is a prime example of problems encountered in a complex system, such as the supply chain. In these systems, problems are multifaceted with a primary problem and many related sub-problems. For instance, the bullwhip effect, one of the fundamental problems in supply chain management literature (Lee et al. 1997a), has several secondary problems, such as order management, demand forecasting management, inventory management, and shipment consolidation. Further, Lee et al. (1997a) formally identify the main causes of the bullwhip effect, while Lee et al. (1997b) discuss their managerial implications. They state that if the following conditions hold—(1) demand is mean stationary and no signal processing

is used, (2) lead time is zero, (3) fixed ordering cost is zero, and (4) no price variation occurs—then the order variance increase does not occur. However, if some of these conditions are relaxed, the bullwhip effect may be observed.

Classification of Techniques
We need to discuss some of the published techniques utilized in managing the bullwhip effect to highlight their classification. Chen et al. (2000a) use the simple moving average forecasting technique to obtain forecasts and investigate the bullwhip effect according to lead time and information sharing. Chen et al. (2000b) and Xu et al. (2001) use the exponential smoothing technique in forecasting. Chen et al. (2000b) also show that, if a smoothing parameter in exponential smoothing is set to have equal forecasting accuracy for both exponential smoothing and moving average methods, then exponential smoothing gives larger order variance. Graves (1999) demonstrates the presence of the bullwhip effect, if external demand, which is the first-order integrated moving average process, is forecasted using exponential smoothing with an optimally set smoothing parameter. Metters (1997) measures the impact of the bullwhip effect by comparing results obtained for highly variable and seasonal demand against the case with low demand variability and weak seasonality. Cachon (1999) proposes methods to reduce the bullwhip effect using balanced ordering.

Problem model taxonomy is a projection of system taxonomy, and thus inherits system structure and vocabulary. A problem domain is presented at two levels—generic problem domain and specific problem domain. Generic problem domain taxonomy is a class of problems that can occur in a supply chain, such as *coordination of production activities*. It is a highly generic problem that comprises several tasks, such as *scheduling of production or inventory replenishment*, which are problems describing more specific issues. Usually, specific problem domain taxonomy is represented by domain-dependent (or specialized) model(s). Splitting problem representation modeling into the above defined two parts provides the means for developing generic and specific problem models.

The process of problem model taxonomy development starts with problem domain space identification. This involves analysis and design of functional requirements for the problem and proposing a structured representation of relevant information. For example, for *a scheduling of production problem*, the model comprises its input and output variables, underlying sub-tasks or activities, tools and mechanisms for solving the problem, problem-oriented goals, roles and agents involved in performing them in accomplishing tasks to achieve identified goals, and external environmental issues. The purpose of problem taxonomy (PT) is the systematic representation of supply chain domain constituents, such as problems and their content.

Different problem models have the same representation format and characteristics vocabulary, thus providing standardization of information representation in the supply chain domain. Problem model taxonomy serves as a meta-model for knowledge model generation and ontology engineering. Ontology inherits concepts, subsumption relationships, and characteristics from the problem model, thus providing consistency in representing various problems. Ontology development components enrich the problem model with constructs, thus turning it from an abstract problem representation into a knowledge model by formulating rules and regulations related to the problem domain. These constructs are: (1) axioms, defining rules specific to the problem domain; (2) algorithms, providing step-by-step procedures for approaching the problem and solving it; and (3) commitments, linking characteristics to data and assigning variables with values. The first two components are modeled through a comprehensive analysis of the problem. System analysis and design techniques, such as process modeling and object-oriented design, are applied for this purpose. The identification of the first two components is the most important part of ontology development. Ontology by itself is a vocabulary with rules on its use. Real world applications require data to operate. Ontological commitments provide these data.

Object model generation is a software engineering practice. If parallels are drawn with software engineering, ontologies can be considered as classes, while object models are their instances encapsulated into software entities. Object models are tangible software constructs, where problem-specific data are represented in a common programming language, encapsulated in a formal model, and accompanied with descriptions of what to do with the data and how to do it.

The next section formalizes the proposed framework with the help of a knowledge management system reference model.

6.5   Knowledge Management System Reference Model

The knowledge management system reference model is proposed as a theoretical foundation for building knowledge-based information systems. It follows various stages in the above-described framework and formally represents its component types, their meaning, and functions. The reference model is divided into three parts: source system representation (system taxonomy), supply chain functional requirements representation (problem taxonomy), and formal knowledge representation (ontology). First, notation to represent the reference model is presented. Next, the reference model is formally enumerated in the form of a set of equations.

Notations related to general problem representation	
S	System
T	Thing symbolizing the elements of a system
R	Relationships among things of a system defined on T
GP	Generic problem model
at_i	Attribute (the index i here and afterwards signifies the ith attribute in the set of attributes)
At_i	Set of instances of at_i attribute
vv_i	Variable that can be assigned to attribute at_i for generic problems
VV_i	Set of possible values that variable vv_i may possess

(continued)

ww_i	System generic state for vv_i, respectively
WW_i	Set of possible states of ww_i
Notations related to specific problem representation	
Ob	Object model
b_i	Observation channel
B_i	Set of possible b_i states
SP	Specific problem model
v_i	Variable that can be assigned to attribute at_i for specific problems
V_i	Set of possible values that variable v_i may possess
w_i	System specific state
W_i	Set of possible states w_i
o_i	Observation channel for attributes at_i
$\bar{O}$	Relationship between object system and problem system
W	Class instances of S for supply chain domain (general representation of W_i)
Notations common for specific and general problem representations	
$\hat{E}$	Relationship between specific and generic systems
e_i	Relationship between V_i, VV_i
k_j	Relationship between W_j, WW_j (the index j signifies the jth relationship between general WW_j and specific W_j system states as well as between B_j and W_j)
S_w	Specific system for supply chain domain (an instance of S)
T_w	Things specific to supply chain domain (an instance of T)
R_w	Set of relationships held on T_W
Notations for ontology	
M	Data model for a supply chain domain
I	Ontological commitments
V	Set of variables (General representation of V_i)
B_w	Observation channels for defining variables
B_C	Observation channels for defining constraints
B_H	Observation channels for defining algorithms
J	Set of Interpretation functions I
M_w	Data model for a supply chain problem
C	Constraints on data
O	Ontology model
A	Set of axioms
H	Algorithm or heuristics
G	Set of equations

The system consists of interrelated elements. Ackoff (1971) identifies system characteristics, such as an abstract and a concrete system, system state, its changes, and so on. These characteristics guided us during the development of the reference model. System taxonomy is an abstract system whose elements are concepts. Problem taxonomy has two system representation forms—abstract system representation, where problem-relevant elements are presented as concepts: and concrete system representation, where these elements are presented as objects. Ontology presents the states of the system from both static and dynamic perspectives. Ontology also presents system behavior such as response or reaction.

The approach for source abstract system representation is adopted from Klir (1984) and can be formulated as follows:

$$S = (T, R) \tag{6.1}$$

Formally, a supply chain system can be represented as a collection of all possible instances of a generic system applied to a supply chain, with corresponding relationships

$$S = (T, W, R) \tag{6.2}$$

Equation (6.1) is a highly generic and domain-independent system representation. Equation (6.2) is still generic, but is a domain-dependent representation. Only those things and their relationships are considered to exist in system instances W. Equation (6.2) is the system taxonomy formalism, where W is the supply chain domain. A detailed description of the taxonomy of a system in general can be found in Chandra et al. (2007). For each possible system instance $w \in W$, the intended structure of w according to S is the structure (problem classification in problem taxonomy)

$$S_w = (T_w, R_w) \tag{6.3}$$

R_w is the set of extensions (relative to w) of elements of T_w

$$R = \{R_w | w \in W\}, T = \{T_w | w \in W\} | \tag{6.4}$$

We denote with S the set of all the intended system instance structures of the system

$$S = \{S_w | w \in W\} \tag{6.5}$$

Equations (6.4) and (6.5) reveal that for each system instance w, there is only one system structure S_w with one set of things T_w and one set of relationships R_w. Each S_w is a description of a problem (model) defined as a part of problem taxonomy for which a solution is to be found. S_w contains the names of parameters identified in the problem description, with corresponding relationships organized in a structured hierarchy. Problem model development based on this formalism offers two sub-levels of the problem-modeling layer: problem object model and problem formal model.

Problem Object Model and Problem Formal Model
The notion of thing is abstract. To investigate a single thing, we separate it from the outside world, and examine it as an object.

$$Ob = \left(\left\{ (at_i, At_i) \,\middle|\, i \in N_n \right\}, \left\{ (b_j, B_j) \,\middle|\, j \in N_m \right\} \right) \tag{6.6}$$

where $N_n = \{1,2,\ldots,n\}$ is the number of attributes that the object Ob possesses; and $N_m = \{1,2,\ldots,m\}$ is the number of observation channels where attributes are examined and collected. N_n and N_m are the rows and columns, respectively, of a two-dimensional matrix with n rows and m columns. Observation channels are situations, circumstances, processes, narrative descriptions, or any other sources where the problem can be investigated. The set of possible observation channels is denoted by B_j. Different observations where attributes are examined are called backdrops. When investigating a more specific system, backdrops can be considered as situations, where we examine the same attribute. These situations can be subdomains or problems. (at_i, At_i) denotes an attribute and a set of its appearances (possible values that the attribute can possess), respectively. (b_j, B_j) denotes an observation channel and a set of its states, respectively.

Ob is the object (an instance of a thing). Variables are used for an operational representation of an attribute. Each attribute has a name, which is taken from the set of possible values (At_i).

Attributes define two types of variables general and specific for use in general and specific models, respectively. General and specific variables are components of three primitive systems: object system, specific problem system, and general problem system. The last two primitive system representations connect observed domain attributes to real world variables, which this book classifies as ontological commitments. Separation of generic and specific objects is comparative. In some situations, only one problem model is required, while in other cases two or more problem models are necessary to alleviate the complexity by separating problem domains into information models with various levels of abstractions. Two levels of abstraction are discussed: generic Eq. (6.8) and specific Eq. (6.7).

$$SP = \left(\left\{ (v_i, V_i) \,\middle|\, i \in N_n \right\}, \left\{ (w_j, W_j) \,\middle|\, j \in N_m \right\} \right) \tag{6.7}$$

$$GP = \left(\left\{ (vv_i, VV_i) \,\middle|\, i \in N_n \right\}, \left\{ (ww_j, WW_j) \,\middle|\, j \in N_m \right\} \right) \tag{6.8}$$

SP contains variables v_i related to a specific problem, and GP contains variables related to a general problem vv_i. Both specific and general variables may have sets of states (values vv_i; VV_i) and participate in a set of system states (situations ww_j; WW_j). A specific problem model contains variables related to a set of abstractions (one for each variable), expressing the relationships between specific and general problem systems. It can be called an abstraction channel, which formally can be represented in Eq. (6.9) as

$$\hat{E} = \left(\left\{ (VV_i, V_i, e_i) \,\middle|\, i \in N_n \right\}, \left\{ (WW_j, W_j, k_j) \,\middle|\, j \in N_m \right\} \right) \tag{6.9}$$

The relationship between the object system and the problem system Eq. (6.10) is expressed by an observation channel consisting of individual observation channels for each attribute in the examined system.

$$\widetilde{O} = \left(\left\{ (At_i, V_i, o_i) \big| i \in N_n \right\}, \left\{ (B_j, W_j, w_j) \big| j \in N_m \right\} \right) \tag{6.10}$$

The notion of thing about a particular problem can be formulated as

$$T_w = (Ob, GP, SP) \tag{6.11}$$

Relationships among things can be formulated as

$$R_w \cup \widetilde{O} \cup \widehat{E} \tag{6.12}$$

Equation (6.12) comprises all possible relationships that may exist in system instance w. To keep the model simple, we will refer to the problem model Eq. (6.3) as the object model and to the set of relationships as R_w. A problem model S_w is an abstract representation of a problem domain—a meta-model. Ontological commitments are for developing a data model out of this meta-model. These commitments are interfaces between abstract problem representation and real world data storage. Rearranging the standard definition, we can define a model M as a structure (S, I), where $S = (T, R)$ is a global structure (standard system definition) and I is an interpretation function assigning elements of T to constant symbols (variables) of V.

$$M = (S, I) \tag{6.13}$$

$$I = (V \rightarrow T_w \cup B_w) \tag{6.14}$$

I in Eqs. (6.13) and (6.14) is a function that through observation channel B assigns attributes of T to variables of V. This intentional interpretation can be classified as the first ontological commitment. Observation channels are situations where the system state is captured to observe variables V. These can be process models, where required variables participate. Studying the documented process model may reveal the meaning of variables and where they can be taken from. Another example of an observation channel can be database schema, precisely describing how variables can be queried and stored. The model representation can be used for more general cases:

$$M = (T, W, R, J) \tag{6.15}$$

These are data models for a system in general, including all of its instances and possible interpretations. Equation (6.15) is not practical, because it will never be implemented for presenting actual system data models. Rather, data model presentations for specific system instances are more practical. If we assume $S_w \in S$, for each instance $w \in W$

$$M_w = (T_w, R_w, I) \tag{6.16}$$

M_w is a projection of M, but we refer to it as a model, not a model instance, because model M in reality will never be implemented. A model can describe a situation common to many states. The second ontological commitment is the application of logical axioms designed to account for the intended meaning of vocabulary, and assigning constraints to system variables. Ontology can be represented as a continuation of problem representation by adding new features to it.

$$O = (M, A, H) \tag{6.17}$$

A is a set of axioms for assigning constraint C to variables through the B_C observation channel.

$$A = (C \rightarrow V \cup B_C) \tag{6.18}$$

H is a set of algorithms for assigning mechanisms (G) to data model processing through the B_H observation channel.

$$H = (G \rightarrow M \cup B_H) \tag{6.19}$$

6.6 Development of Components of Knowledge Management System

Ontology development is the implementation of the reference model described in the previous section in capturing its elements, assembling them using a computational language, and storing them in an environment that would facilitate dissemination and usage. Particularly, software tools and techniques will use the developed ontology as part of supply chain decision modeling, the other significant part of supply chain configuration, which is taken up for discussion in Chaps. 7–11 of this book. Various stages of ontology development are described next.

6.6.1 Capture

This stage involves the following activities: (1) identification of key concepts and relationships in the domain of interest, (2) production of unambiguous text definitions for such concepts and relationships, and (3) identification of terms to refer to such concepts and relationships (Uschold and Gruninger 1996). Development of a system taxonomy aims to achieve the first two activities for the supply chain domain in general. Identification of concepts for a specific purpose and scope (i.e., the third activity), is the task for the ontology capture activity. The difference between ontology development from scratch and using system taxonomy is that the latter uses search and navigation in the taxonomy hierarchy to find relevant concepts. Once concepts are chosen, an instance of system taxonomy is created that captures only selected concepts, which is T_w and specified by Eq. (6.11). As was mentioned earlier, for the sake of simplicity, thing T_w will be regarded as the object model Eq. (6.6). Relationships R_w are captured automatically in the form of a taxonomy structure that defines how concepts relate to each other.

The knowledge management system conceptualization framework, in addition to the data model (M) identified in the previous section, defines two other components—axioms (A in Eq. (6.18)) and algorithms (H in Eq. (6.19)). Axioms and algorithms capture a process for a search of rules held in the domain of interest for which an ontology is to be built. The theory for axiom representation is based on situation calculus and predicate calculus for representing a dynamically changing supply chain environment. Situation theory (Lesperance et al. 1995) views a domain as having a state (or situation). When the state is changed, there is a necessity for an action. Predicate theory defines conditions on which specific actions can be taken. Based on these two theories, ontology calculus for a supply chain is planned to be built. It will be based on extending both predicate and situation calculus with new terminology specific to the supply chain domain. The term *do(x, s)* represents the state after an agent performs an action *x* in state *s*. A more supply chain-specific example can be the statement that each product should have demand. This can be formulated as *Exist (demand, Product)*. Another example of an axiom is the inventory constraint: Maximum Inventory $\geq$ Current Inventory Level, which can be formulated as *Less(MaxInventory, CurrInventory)*.

An example of a portion of an algorithm is the formula according to which order size is calculated as $s = L \times AVG + z \times STD$; IF IL $< s$ THEN Order $=$ s-IL, where *s* is the reorder level, *L* is lead time; *AVG, STD* are forecasted demand means and standard deviation, respectively, and *z* is a customer service indicator. If the inventory level (IL) is less than the calculated reorder level, an order is placed (Order), which is equal to the difference of reorder and inventory levels. This axiom can be formulated through situation calculus as $Poss(do((L \times AVG + z \times STD) = s) > Il) = MakeOrder(s - Il)$.

6.6.2 Assembly

Assembly is an explicit representation in some formal language of the conceptualization captured in the preceding stage. This involves (1) committing to the basic terms that will be used to specify the ontology, (2) choosing a representation language, and (3) writing the code. It simply has to do with writing down, in some language or communicative medium, descriptions or pictures that correspond in some salient way to the world, or a state of the world, of structured data.

For ontology representation, different programming languages and standards have been utilized. Ontolingua (Farquhar et al. 1997) adds primitives to defined classes, functions, and instances. Ontolingua is not a representation system, but rather a mechanism for translating from standard syntax to multiple representation systems. OIL (Ontology Interchange Language) (Fensel et al. 2002) fuses two paradigms—frame-based modeling with semantics based on description logic, and syntax based on web standards, such as extensible markup language (XML) schema and resource description framework (RDF) schema. Both Ontolingua and OIL are frame-based languages that do not provide formalism for first-order logic. With the latter, we intend to represent process logic the same way as frame logic.

XML has become a standard for communication between heterogeneous systems and is widely used on the Internet (Staab et al. 2001). This presents new opportunities for knowledge representation and acquisition and has two aspects. First, XML documents can easily be translated into knowledge representation format and parsed by problem-solving environments or domains. Second, XML can directly connect with data storage repositories (RDBMS or ERP systems), thus enabling database queries to be more expressive, accurate, and powerful. The two objectives can be achieved by enhancing the semantic expressiveness of XML, especially XML data schemas (XSD). We propose a new language, Supply Chain Markup Language (SCML), for presenting knowledge about supply chains. The specification of SCML is formulated as a XSD data schema, depicted in Fig. 6.2. It reflects system representation formalism presented in system taxonomy. At the top level there are seven groupings: input, output, functions, environment, processes, and mechanisms. Each grouping is a container, which consists of subclasses.

A representative sample of SCML is depicted in Fig. 6.3. It defines the entity Axioms, any elements it may have, and entities it may contain. An Axioms entity class may have one or many Rules (unbounded) entities, which may have Attributes entities (0 or many). An Argument entity may have two attributes: Name and Description. The entity Rule may have one and only one Body entity and two attributes.

The SCML XSD specification defines the format of knowledge representation and can be used for developing ontology models and verifying their correctness.

The assembly process, as viewed in this chapter, is the representation of captured knowledge with XML formalism. Three components of an ontology model can be represented. Data assembly is concerned with developing software programs for connecting to data storage facilities and building XML data files based on schema described earlier. A data model example is represented in Fig. 6.4. Demand for a part (Number 295) produced for Customer Number 21 is demonstrated with four attributes. The Demand Net attribute can be used if the demand is stationary. In case it is dynamic, the DemandMeans and DemandDeviation pair of attributes can be used to present the demand distribution function (it is assumed that demand has a normal distribution). The numberofRegression attribute presents the number of observations when calculating its mean and deviation.

Axiom assembly is a manual process consisting of manually entering rules captured with ontology calculus into an XML data file based on SCML schema.

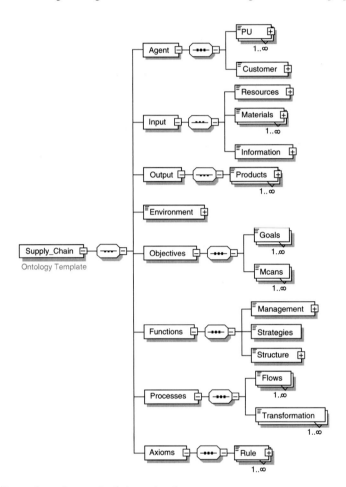

Fig. 6.2 Data schema for supply chain markup language

An axiom model XML example is represented in Fig. 6.5. The rule demonstrated here is about the relationship between inventory and demand, in case the service level is 100 %. Ontology calculus for this rule looks like:

$$Poss(ServiceLevel = 100\%) = Less(CurrInventory, Demand)$$

Ontology calculus formalism is transformed into XML formalism as follows. The entity type Rule defines the condition Service level is 100 %. It contains two arguments, inventory and demand. The entity Body defines the relationship between these two arguments according to the condition identified in the parent Rule entity's Name property.

```
<xs:element name="Axioms">
    <xs:complexType>
        <xs:sequence>
            <xs:element name="Rule" maxOccurs="unbounded">
                <xs:complexType>
                    <xs:sequence>
                        <xs:element name="Body"/>
                        <xs:element name="Attribute" minOccurs="0" maxOccurs="unbounded">
                            <xs:complexType>
                                <xs:attribute name="Name" type="xs:string" use="optional"/>
                                <xs:attribute name="description" type="xs:string"
                                        use="optional"/>
                            </xs:complexType>
                        </xs:element>
                    </xs:sequence>
                    <xs:attribute name="Number" type="xs:string" use="optional"/>
                    <xs:attribute name="Name" type="xs:string" use="optional"/>
                </xs:complexType>
            </xs:element>
        </xs:sequence>
        <xs:attribute name="Name" type="xs:string" use="optional"/>
        <xs:attribute name="ID" type="xs:string" use="optional"/>
    </xs:complexType>
</xs:element>
```

Fig. 6.3 Supply chain markup language example axioms

Fig. 6.4 Data model XML
fragment

```
<Demand>
    <Attribute part="295" customer="21"
    DemandNet="5984"
    DemandMeans="6868"
    DemandDeviation="1000"
    numberofRegression="20"
</Demand>
```

```
<SupplyChain>
    <Axioms>
        <Rule Number="1" Name="Service level is 100%">
            <Argument Name="Inv" Description="Inventory"/>
            <Argument Name="Dm" Description="Demand"/>
            <Body>Inv&gt;=Dm</Body>
        </Axioms>
</SupplyChain>
```

Fig. 6.5 Axiom model XML fragment

6.6.3   Storage

The purpose of building an ontology server is to enable technology that will
facilitate the large-scale reuse of ontologies through Web interfaces for decision-

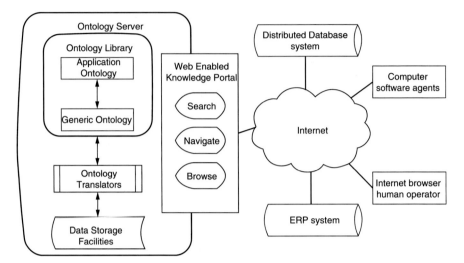

Fig. 6.6 Ontology Server architecture

making purposes throughout the complex, extended supply chain enterprise. Figure 6.6 depicts the ontology server architecture. Data are stored in data storage facilities, which are mainly relational database systems or other carriers of information, such as ERP repositories (as a complex system for maintaining data) or files (as a simpler class of data storage facilities).

The main component of the ontology server is the ontology library, which must have an index indicating how each individual item can be found. A problem classification hierarchy, Eq. (6.4), defines this index. Each node in this library corresponds to a problem and is the ontology specifying description of that problem.

6.6.4   Usage

Ontology can be utilized in a variety of ways. It can serve as an explicit medium where knowledge workers share their expertise and skills, it can be used as specifications for software engineers in developing complex software applications, and it can be used by decision makers for understanding the problem and making decisions. But the greatest advantage of having explicit ontologies is in implementing the vision of supply chain management as formulated by Fox et al. (2000). According to Fox, a supply chain is viewed as being managed by a set of intelligent agents, each responsible for one or more tasks in the supply chain, and each interacting with other agents in the planning and execution of their responsibilities.

An ontology server deployed on the Web makes the library available to supply chain members, who share the same perception of problems that they communicate to each other. In the case of agent-based SCM, ontologies provide the members with communication and interoperation. Ontologies inform the system user of the vocabulary for representing domain or problem knowledge.

Marra et al. (2012) identify that outsourcing, new product development, decision support, and risk management all are relevant to supply chain configuration. The knowledge management framework is particularly well-suited for addressing decision-making challenges in a distributed environment. Liu et al. (2014) argue that increasingly supply chain knowledge management systems should incorporate global contextual information and knowledge.

6.7 Summary

In this chapter, we explore the complexity of the supply chain configuration problem and argue that the best way to solve it is through devising a crosscutting approach that adopts concepts drawn from various disciplines in designing, developing, and implementing efficient and effective solutions. We take the representative information integration problem in the supply chain and argue that any methodology developed for supply chain configuration must explicitly take into account the systemic, reductionist, and analytic approaches available either in the published literature, or designed specifically. These approaches incorporate supply chain configuration problem details at the abstract, activity, and implementation levels, respectively. We make a case for knowledge design, development, and dissemination using taxonomy and ontology principles to incorporate system integration concepts. We propose a theoretical knowledge management system development framework, which acts as a reference model. It is based on problem solving at the above three levels. We also describe a brief implementation scenario of this framework. The utility of this framework rests on the fact that it provides a high level approach to managing the generated supply chain problem-solving knowledge, using various techniques described in Chaps. 7–9.

References

Ackoff RL (1971) Towards a system of systems concepts. Manag Sci 17:661–671

Bellamy MA, Basole RC (2013) Network analysis of supply chain systems: a systematic review and future research. Syst Eng 16:235–249

Blanchard SB, Fabrycky WJ (1990) Systems engineering and analysis. Prentice Hall, Englewood Cliffs, NJ

Cachon GP (1999) Managing supply chain demand variability with scheduled ordering policies. Manag Sci 45:843–856

Chandra C (1997) Enterprise architectural framework for supply-chain integration. Proceedings of the 6th annual industrial engineering research conference, Miami Beach, USA, pp. 873–878, Accessed 17–18 May

Chandra C, Grabis J, Tumanyan A (2007) Problem taxonomy: a step towards effective information sharing in supply chain management. Int J Prod Res 45:2507–2544

Chen F, Drezner Z, Ryan JK, Simchi-Levi D (2000a) Quantifying the bullwhip effect in a simple supply chain: the impact of forecasting lead times and information. Manag Sci 46:436–443

Chen F, Ryan JK, Simchi-Levi D (2000b) Impact of exponential smoothing forecasts on the bullwhip effect. Nav Res Logist 47:269–286

Clark AJ (1972) An informal survey of multi-echelon inventory theory. Nav Res Logist Q 19:621–650

Clark AJ, Scarf H (1960) Optimal policies for multi-echelon inventory problem. Manag Sci 6:475–490

Cohen MA, Lee HL (1989) Resource deployment analysis of global manufacturing and distribution networks. J Manuf Oper Manag 2:81–104

Deleersnyder JL, Hodgson TJ, King RE, O'Grady PJ, Savva A (1992) Integrating kanban type pull systems and MRP type push systems: insights from a Markovian model. IIE Trans 24:43–56

Diks EB, De Kok AG (1998) Optimal control of a divergent multi-echelon inventory system. Eur J Oper Res 111:75–97

Diks EB, De Kok AG, Lagodimos AG (1996) Multi-echelon systems: a service measure perspective. Eur J Oper Res 95:241–263

Drew SAA (1975) The application of hierarchical control methods to a managerial problem. Int J Syst Sci 6:371–395

Farquhar A, Fikes R, Rice J (1997) The ontolingua server: a tool for collaborative ontology construction. Int J Hum Comput Stud 46:707–727

Fensel D, Harmelen F, Horrocks I, McGuinness DL, Patel-Schnaider PF (2001) OIL: an ontology infrastructure for semantic web. IEEE Intell Syst 2:38–45

Fensel D, Hendler J, Lieberman H, Wahlster W (2002) On-to-knowledge: semantic web enabled knowledge management semantic web technology. MIT Press, Boston, MA

Fox MS, Gruninger M (1999) Ontologies for enterprise integration. Department of Industrial Engineering, University of Ontario, Oshawa, ON

Fox MS, Barbuceanu M, Teigen R (2000) Agent-oriented supply-chain management. Int J Flex Manuf Syst 12:165–188

Graves SC (1982) Using Lagrangean techniques to solve hierarchical production planning problems. Manag Sci 28:260–275

Graves SC (1999) A single-item inventory model for a nonstationary demand process. Manuf Serv Oper Manag 1:50–61

Gruber TR (1993) A translation approach to portable ontologies. Knowl Acquis 5:199–220

Gruber TR (1995) Toward principles for the design of ontologies used for knowledge sharing. Int J Hum Comput Stud 43:907–928

Grubic T, Fan I (2010) Supply chain ontology: review, analysis and synthesis. Comput Ind 61:776–786

Grubic T, Veza I, Bilic B (2011) Integrating process and ontology to support supply chain modelling. Int J Comput Integrated Manuf 24:847–863

Gruninger M (1997) Integrated ontologies for enterprise modelling. In: Kosanke K, Nel J (eds) Enterprise engineering and integration building international consensus. Springer, Berlin, pp 368–377

Gruninger M, Atefi K, Fox MS (2000) Ontologies to support process integration in enterprise engineering. Comput Math Organ Theor 6:381–394

Guarino N (1995) Formal ontology conceptual analysis and knowledge representation. Int J Hum Comput Stud 43:625–640

Gupta S, Loulou R (1998) Process innovation, product differentiation and channel structure: strategic incentives in a duopoly. Market Sci 17:301–316

Hackman ST, Leachman RC (1989) A general framework for modeling production. Manag Sci 35:478–495

Hariharan R, Zipkin P (1995) Customer-order information leadtimes and inventories. Manag Sci 41:1599–1607

Hirsch B (1995) Information system concept for the management of distributed production. Comput Ind 26:229–241

Huang C, Lin S (2010) Sharing knowledge in a supply chain using the semantic web. Expert Syst Appl 37:3145–3161

IMTR (1999) IMTR Technologies for Enterprise Integration, Rev 31; Integrated Manufacturing Technology Roadmapping Project. Oak Ridge Centers for Manufacturing Technology, Oak Ridge, Tennessee

ISO TC 184/SC 5/WG 1 (1997) Requirements for enterprise reference architectures and methodologies. http: //www.melnistgov/sc5wg1/gera-std/ger-anxshtml

Jennings R (1994) Cooperation in industrial multi-agent systems, vol 43, World scientific series in computer science. World Scientific Publishing Co., Singapore

Klir GJ (1984) Architecture of systems problems solving. Plenum Press, New York

Klir GJ (1991) Facets of systems science. Plenum Press, New York

Kosanke K (1995) CIMOSA – overview and status. Comput Ind 27:101–109

Lambert DM, Cooper MC, Pagh JD (1998) Supply chain management: implementation issues and research opportunities. Int J Logist Manag 9:1–19

Lee HL (1993) Effective inventory and service management through product and process redesign. Oper Res 44:151–159

Lee HL, Billington C (1993) Material management in decentralized supply chains. Oper Res 41:835–847

Lee HL, Padmanabhan V, Whang S (1997a) The bullwhip effect in supply chains. Sloan Manage Rev 38:93–102

Lee HL, Padmanabhan V, Whang S (1997b) Information distortion in a supply chain: the bullwhip effect. Manag Sci 43:546

Lesperance Y, Levesque HJ, Lin F, Scherl RB (1995) Ability and knowing how in the situation calculus. Stud Logica 66:165–186

Little JDC (1992) Tautologies models and theories: can we find "laws" of manufacturing? IIE Trans 24:7–13

Liu S, Moizer J, Megicks P, Kasturiratne D, Jayawickrama U (2014) A knowledge chain management framework to support integrated decisions in global supply chains. Prod Plann Contr 25:639–649

Malone TW, Crowston K (1994) The interdisciplinary study of coordination. ACM Comput Surv 26:87–119

Malone TW, Crowston K, Lee J, Pentland B, Dellarocas C, Wyner G, Quimby J, Osborn CS, Bernstain A, Hermen G, Klein M, O'Donnel E (1999) Tools for inventing organizations: towards a handbook of organizational processes. Manag Sci 45:425–443

Marra M, Ho W, Edwards JS (2012) Supply chain knowledge management: a literature review. Expert Syst Appl 39:6103–6110

Masahiko A (1984) The co-operative game theory of the firm. Oxford University Press, Oxford, UK

McCarthy J (2003) What is artificial intelligence? Computer Science Department Stanford University. http://www-formalstanfordedu/jmc/whatisai/whatisaihtml

McKelvey B (1982) Organizational systematics taxonomy evolution classification. University of California Press, Berkeley, CA

Meersman R (2001) Ontologies and databases: more than a fleeting resemblance. Rome OES/SEO Workshop

Metters R (1997) Quantifying the bullwhip effect in supply chains. J Oper Manag 15:89–100

Morris WT (1967) On the art of modeling. Manag Sci 13:B707–B717

NIST (1999) Manufacturing enterprise integration program. National Institute of Standards and Technology, Gaithersburg, MD

Pritsker AAB (1997) Modeling in performance-enhancing processes. Oper Res 45:797–804

Pyke DF, Cohen MA (1990) Push and pull in manufacturing and distribution systems. J Oper Manag 9:24–43

Shaw MJ, Solberg JJ, Woo TC (1992) System integration in intelligent manufacturing: an introduction. IIE Trans 24:2–6

Sousa P, Heikkila T, Kollingbaum M, Valckenaers P (1999) Aspects of cooperation. Distributed manufacturing systems. Proceedings of the second international workshop on intelligent manufacturing systems, pp 685–717

Sowa J (2000) Ontology metadata and semiotics. International conference on conceptual structures ICCS'2000, Darmstadt, Germany, pp 4–18

Staab S, Schnurr HP, Studer R, Sure Y (2001) Knowledge processes and ontologies. IEEE Intell Syst 16:26–34

Stader J (1996) Results of the enterprise project. Proceedings of expert systems 1996 the 16th annual conference of the British Computer Society Specialist Group on Expert Systems, Cambridge, UK

Stumptner M (1997) An overview of knowledge-based configuration. AI Comm 10:111–125

Swaminathan JM, Smith SF, Sadeh NM (1998) Modeling supply chain dynamics: a multiagent approach. Decis Sci 29:607

Thomas DJ, Griffin PM (1996) Coordinated supply chain management. Eur J Oper Res 94:1–15

Tzafestas S, Kapsiotis G (1994) Coordinated control of manufacturing/supply chains using multi-level techniques. Comput Integrated Manuf Syst 7:206–212

Uschold M, Gruninger M (1996) Ontologies: principles methods and applications. Knowl Eng Rev 11:93–155

Whang S (1995) Coordination in operations: a taxonomy. J Oper Manag 12:413–422

Wooldridge M, Jennings NR (1995) Intelligent agents: theory and practice. Knowl Eng Rev 10:115–152

Xu K, Dong Y, Evers PT (2001) Towards better coordination of the supply chain. Transport Res E Logist 37:35–54

Younis MA, Mahmoud MS (1986) Optimal inventory for unpredicted production capacity and raw material supply. Large Scale Syst 11:1–17

Chapter 7
Conceptual Modeling Approaches

7.1 Introduction

An understanding of information flows and processing functions is essential for any decision-modeling effort. Traditionally, these information flows are described in terms that are specific to particular decision- modeling techniques. However, in the heterogeneous supply chain environment, that results in largely diverse and often incompatible data definitions. Therefore, a more unified approach to representing information flows and their processing functions is required. Conceptual modeling techniques, long-used for information systems development are well-suited for these purposes. Supply chain management is an area where interactions between Decision Sciences and Information Systems Engineering are most profound. Implementation of decisions is not possible without information systems support, and information systems alone without decision-making components are no longer sufficient to maintain a competitive advantage.

Conceptual modeling is a key part of the information systems modeling process, where models undergo different phases of elaboration starting with general requirements for information systems, down to semi-executable models directly used in the implementation phase. The main purpose of these models is to simplify development and maintenance complexity of large information systems by describing the system using less abstract concepts. That is especially important for channeling user requirements to system developers. For instance, supply chain process models are shown to have a major impact on ensuring business and information systems alignment (Millet et al. 2009) and common reference models are a key to supply chain integration (Chan and Kumar 2014).

Similarly, conceptual and information modeling can be used to describe complex decision-modeling problems (Biswas and Narahari 2004; Kim and Rogers 2005). Besides the descriptive capabilities of information modeling techniques helping to understand the problem, developed information models provide a link between decision-modeling and the enterprise-wide information system.

© Springer Science+Business Media New York 2016
C. Chandra, J. Grabis, *Supply Chain Configuration*,
DOI 10.1007/978-1-4939-3557-4_7

The objective of this chapter is to describe the application of conceptual modeling techniques for supply chain configuration purposes. The general approach is to use well-known conceptual and information modeling techniques that would enable potential model-driven implementation of decision-modeling components.

The following section describes the purpose of the conceptual modeling. Section 7.3 discusses modeling views including goal, process, and concept models. Based on data modeling described and the literature survey presented in Chap. 3, a generic supply chain configuration model is elaborated in Sect. 7.4 while design of case specific models is presented in Sect. 7.5. Utilization of the conceptual modeling as the basis for model integration is discussed in Sect. 7.6. Section 7.7 offers summary and conclusions for the chapter.

7.2 Purpose

One of the main objectives of using conceptual modeling is providing a relatively easily understandable representation of a problem. Several conceptual and information modeling techniques are usually applied to obtain a comprehensive representation of the problem. The choice of techniques and modeling concepts depends upon objectives of the information modeling application. In the framework of supply chain configuration, several objectives can be identified:

- *Description of the modeling problem.* Information modeling methods are used to attain a better understanding of a particular decision modeling problem. They can be especially useful for describing the decision-making environment. This approach is being used in relation to simulation modeling while it is rarely considered in relation to analytical modeling.
- *Implementation of decision-making components.* If decision-making is to be performed routinely, a software application needs to be developed. Information modeling is an essential part of almost any software development project.
- *Definition of links between decision-modeling and other parts of the enterprise-wide information system.* Decision-making models rely on data provided from other parts of the information system and can also use some functions provided by the supply chain information system. Information modeling is used to map data between components and identify available functions.
- *Integration of the decision-making process with the overall information processing system.* This is similar to the second objective, although a decision-making component becomes an integral part of the supply chain information system in this case. The main problem is ensuring that changes made in both decision-making component and supply chain information system are properly represented in related components.

In the case of implementation of decision-making components, information modeling methods are used in a similar manner as in the development of information systems. This approach is mainly applicable if decision- making components are implemented using general purpose programming languages. If a specialized decision-modeling environment and programming languages are used, application of information modeling is not common and the majority of available tools do not support such an approach.

7.3 Modeling Views

There are multiple conceptualization perspectives of the supply chain configuration problem. These different perspectives are captured using several specific modeling views, which are borrowed from widely used enterprise modeling and information systems development methods and 4EM in particular (Sandkuhl et al. 2014). The modeling views deemed relevant for describing the supply chain configuration problem are goal model, concepts model, business process model and actors model (Fig. 7.1). The goal model is used to represent supply chain configuration goals. It provides criteria for evaluation and comparison of different supply chain configurations as well as to identify synergies and contradictions among supply chain partners. The supply chain partners are defined in the actors model. The actors model is important to represent supply chain power structure and relationships among different legal entities involved in the supply chain. The concept model establishes a common definition of terms used in specification of the supply chain configuration problem. The business process model aids definition of constraints relevant to the supply chain configuration problem with focus on structural and temporal dependencies.

The business process model is represented using the BPMN notation.[1] All other models are represented using the UML notation.[2] Utilization of the widely accepted notation helps to narrow gap between supply chain modeling and related enterprise modeling and information systems development activities.

The conceptual modeling is a time consuming task. In order to address this issue, conceptual models for a particular supply chain configuration initiative are developed on the basis of a generic supply chain configuration model (Fig. 7.2). The generic model attempts to capture common properties of the supply chain configuration problem domain while the case specific model describes the given supply chain and its configuration. The generic model is developed as a reusable representation of the supply chain configuration problem. The case specific model is created by reusing concepts from the generic model and is customized according to specific requirements. It is used for qualitative analysis of the supply chain configuration problem and for providing information to other supply chain configuration models, such as simulation and optimization models.

---

[1] http://www.bpmn.org/.

[2] http://www.uml.org.

Fig. 7.1 Supply chain
configuration
conceptualization views

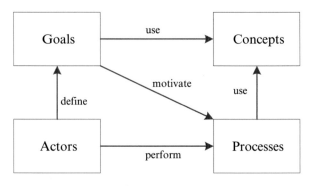

Fig. 7.2 Generic and case
specific models

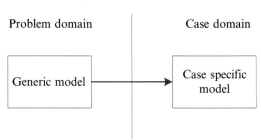

7.4   Generic Model

The generic model is intended as an overall description of the supply chain
configuration problem. It is constructed according to a thorough analysis of supply
chain configuration research reported in Chap. 3 of this book. It captures the most
common features of the supply chain configuration problem in a generalized
manner.

The supply chain configuration goal model (Fig. 7.3) includes three main types
of goals including goals related to supply chain performance, goals related to
definition of supply chain structure and goals related to strategic allocation. The
performance goals are business-oriented goals while the two latter types define
what should be achieved as a result of the supply chain configuration effort. The
performance goals are defined according to the performance attributes defined in
the SCOR model (Stephens 2001; Zhou et al. 2011), and they are common for the
whole supply chain management domain. The structural goals state that the supply
chain units should be selected, their function determined and links among units
should be established as a result of supply chain configuration. The strategic
allocation goals represent the typical allocation decisions made concurrently with
supply chain design.

The supply chain configuration concept model at the generic level is shown in
Fig. 7.4. These concepts are identified according to the literature review reported in
Chap. 3 and represent the most commonly used concepts. From the structural
perspective, it defines types of units and types of links used in supply chains. In

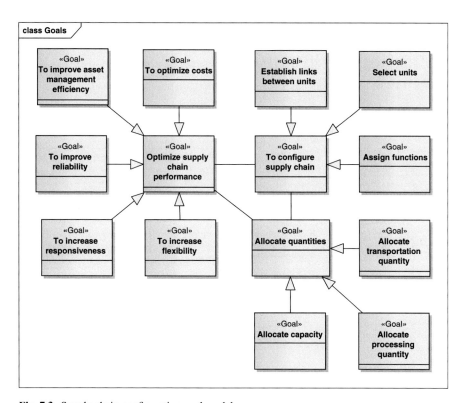

Fig. 7.3 Supply chain configuration goal model

modern supply networks, one unit actually can perform multiple functions. However, the primary function of each unit can still be identified. Additional functions of the unit can be represented using the `Function` concept or more specific Source, Make, Deliver, Return, Plan, and Enable functions, which are defined similarly as in the SCOR model. The `Electronic link` type represents a growing number of products and services transferred digitally. That includes transfer of digital assets such as music and books as well as transfer of product design for digital printing. Similarly, the `Item` subtype `Information` represents such digital services and products rather than the traditional information exchange among supply chain units.

The actor model of the supply chain configuration model shows legal entities involved in the supply chain (Fig. 7.5). Three types of legal entities are identified in the generic model, namely, supply chain partners, managers and customers. The supply chain partner may oversee multiple supply chain units. The manager represents formal or informal and physical or virtual supply chain management center, where decisions concerning supply chain configuration and other supply chain management issues are made (Bitran et al. 2007).

At the generic level, the business process model (Fig. 7.6) shows general temporal relationships among supply chain units. The process model represents

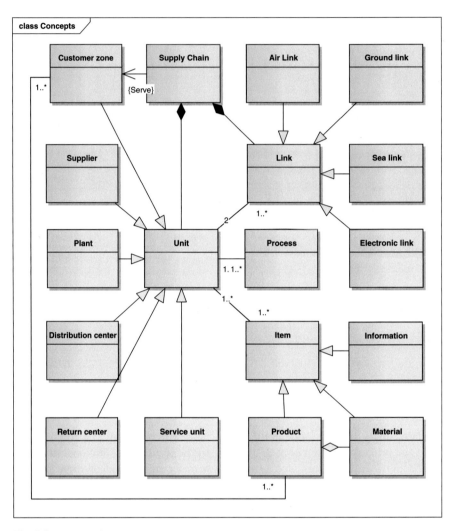

Fig. 7.4 The generic concepts

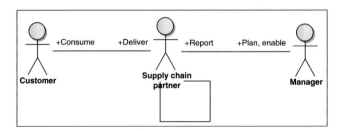

Fig. 7.5 The domain level actor model

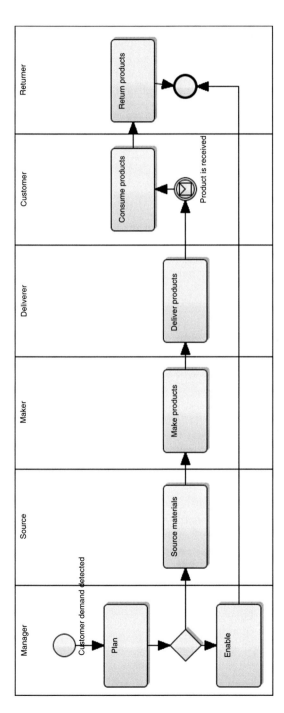

Fig. 7.6 The domain level process model

five main supply chain management processes as defined in the SCOR model. These main processes are plan, source, make, deliver, and return. It is important to note that the SCOR model assumes that these same processes can be used to represent supply chain activities at different levels of abstraction. Thus, Fig. 7.6 shows the process for the whole supply chain while the same processes can be used to model processes at individual supply chain units. The domain level business process model highlight role of manager, who provides plans for supply chain activities and enables actual execution of these activities.

The importance of the business process model increases in case level models.

7.5   Case Specific Model

The case specific model is constructed for a specific supply chain configuration initiative and it uses elements from the generic supply chain configuration model. Development of the case specific model is illustrated using the running example of the SCC Bike supply chain.

7.5.1   Development Principles

The case specific model is intended for analysis of the particular supply chain problem and also serves as input to development of other supply chain configuration models. Depending on the purpose, it has different levels of detail:

- Type level.
- Type level including the most significant instances.
- Instance level.

The type level representation contains only general concepts such as plant, distribution center and product characteristic to the particular supply chain configuration case. These concepts are either selected from the generic model or created anew. The type level model is the most useful for model integration purposes. The instance level representation contains all individual instances of the supply chain network, e.g., individual supply chain units, links, products, and other elements. The instance level representation is the most useful for descriptive analysis of the supply chain network. Given that the supply chain network can be large, often only the most significant instances should be represented individually. In this case, a combination of the type level representation and the instance level representation is used.

7.5.2   *Example*

The case specific supply chain configuration model is created for the SCC Bike supply chain.

The goal, concepts and process models are developed for the case. Figure 7.7 shows the case level goal model, which includes only goals that are the most relevant to the company's supply chain configuration. Profit maximization is the main goal. Additionally, general supply chain problems to be addressed in the SCC Bike case are improvement of responsiveness and increasing flexibility. The responsiveness issue is viewed as an ability to deal with demand variability and the supply chain should be able to meet demand spikes. The flexibility issue is addressed by entering new markets making the supply chain less susceptible to local demand variations. The specific supply chain configuration goals are to select suppliers out of the set of alternative suppliers, to allocation production and distribution quantities to supply chain units as well as to establish links among the units. Addressing the specific problems contributes to improving profitability. The ability to handle demand variations helps to increase sales though it might result in increasing inventory costs.

The concept model at the case level (Fig. 7.8) contains two main types of elements: (1) relevant domain level concepts; and (2) case specific instances of the domain level concepts. The domain level concepts included in the concept model are `Supplier`, `Material`, `Product`, `Customer Zone`, `Distribution Center`, and `Plant`, while concepts such as `Process` are not included because they are not relevant to the particular analysis. The model specifically shows that sea and ground transportation is used to link the indicated type of supply chain units. Associations among these concepts are used to define general relationships among different constituent parts of the supply chain. These associations are subsequently important to define data structure and analysis models (see Sect. 4.3). The `Link` concept is used to represent connections between supply chain units. For instance, it shows that products are delivered from plants to distribution centers via a given link.

The case specific instances of the domain level represent individual supply chain elements such as particular products or suppliers. The supply chain can include a large number of individual elements and only the most important elements are represented separately. In this case, instances are explicitly modeled to represent assembly plants and the frame factory because one of the supply chain configuration decisions is about establishing links among the plants and distribution centers.

Figure 7.9 shows the case level process model. The process model includes both generic supply chain concepts and specific supply chain objects. The generic concepts are used to show that products are shipped from suppliers to plants. The process model shows that in general products can be shipped to any of the distribution centers, as well as customers can be served from any of the distribution centers. However, at the level of specific supply chain objects, there are particular shipment and delivery restrictions. The model shows that frames from the frame

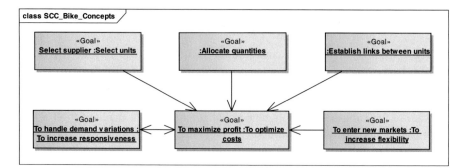

Fig. 7.7 The case level goals

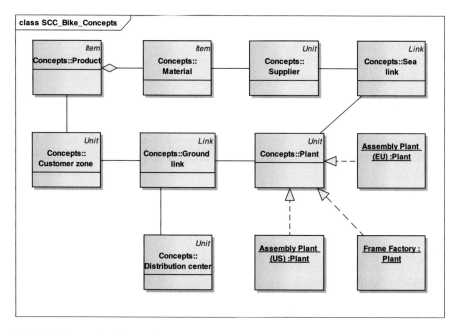

Fig. 7.8 The case level concepts

factory are shipped to other plants. That allows to represent characteristics of the manufacturing process. The process activities correspond to the generic activities used in the SCOR model and include plan, enable, source, make and deliver activities. The model indicates that suppliers engage in delivery activities as far as the SCC Bike is concerned. Plants are responsible for sourcing materials, making products and delivering them to distribution centers, which perform the final delivery to the customers. The process is guided by centralized planning performed at SCC Bike.

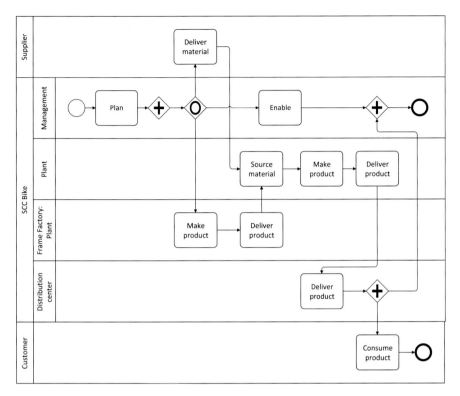

Fig. 7.9   The case level process model

7.6   Model Integration

The conceptual supply chain configuration models can be used on their own right for static inspection and analysis of relationships in the supply chain. However, more importantly these models serve as a common basis for development of utilization of different other qualitative and quantitative models required for comprehensive supply chain configuration decision-making. Having such a common basis reduces the model development effort and ensures consistency among models. Figure 7.10 shows that conceptual models can be used for generation of data structure defining data requirements for solving the supply chain configuration models as well as for development of other supply chain configuration models, such as mathematical optimization and simulation models. The data structure generated is populated by supply chain data and these data are fed into the other supply chain configuration models.

   The data structure is generated from the case level concept model. During the generation process, data structure objects corresponding to concepts defined are created (specific format of data objects depends upon the target implementation platform).

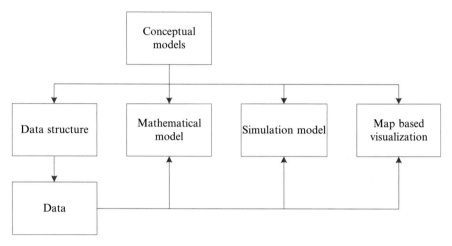

Fig. 7.10 Usage of conceptual models

The modeling technique specific models are developed on the basis of the general data model. Utilization of the general model is useful if multiple specific models need to be constructed. It is also important to enable data mapping between source information systems and decision-modeling components.

In order to generate the data structure from the concept model and to use this data structure for data integration purposes, attributes of concepts should be specified. Attributes can be specified for every element in the concept model and attributes are classified as Decision variables, Parameters, and Metrics (Fig. 7.11). For instance, a plant is described by a parameter capacity.

7.7 Summary

A conceptual representation of the supply chain configuration problem is elaborated by using the enterprise modeling techniques. The configuration problem is modelled at the domain level and at the case level, where the generic concepts are defined at the domain level and instances specific to a particular supply chain are defined at the case specific level. The domain level model is not developed as a meta-model because it can be directly used to create generic supply chain configuration models and other elements can be added, if necessary.

The application of the conceptual models is outlined with emphasis on generation of data structure for storing data necessary for supply chain configuration decision-making. Conceptual models and the generated data structure provide a backbone for more detailed analysis of the supply chain configuration problem using optimization, simulation, and other models. The data structure is generated using the concept model. Other models are important for other types of analysis. For instance, the goal model is important for the AHP to compare different supply

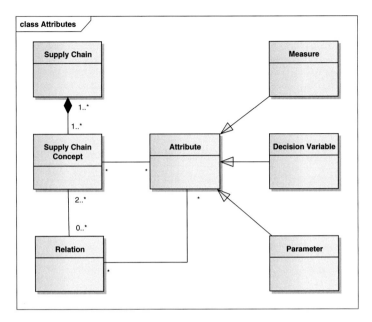

Fig. 7.11 Attributes of the supply chain concepts

chain configuration while the business process model is important in building supply chain simulation models.

The data structure generation is emphasized in the chapter because it is believed that a practical supply chain configuration model at the case level should not include graphical representation of all supply chain entities. The most efficient way is to show graphically only the most important supply chain entities and other supply chain entities can be represented using a text based format. Nevertheless, the textual representation should be easily perceivable and modifiable by experts involved in the supply chain decision-making. Spreadsheets are an attractive option.

References

Biswas S, Narahari Y (2004) Object oriented modeling and decision support for supply chains. Eur J Oper Res 153:704–726

Bitran GR, Gurumurthi S, Sam SL (2007) The need for third-party coordination in supply chain governance. MIT Sloan Manage Rev 48:30–37

Chan HK, Kumar V (2014) Special issue-applications of reference models for supply-chain integration. Prod Plann Contr 25:1059–1064

Kim J, Rogers KJ (2005) An object-oriented approach for building a flexible supply chain model. Int J Phys Distrib Logist Manag 35:481–502

Millet P, Schmitt P, Botta-Genoulaz V (2009) The SCOR model for the alignment of business processes and information systems. Enterprise Inform Syst 3(4):393–407

Sandkuhl K, Stirna J, Persson A, Wißotzki M (2014) Enterprise modeling: tackling business challenges with the 4EM method. Springer, Berlin

Stephens S (2001) Supply chain operations reference model version 5.0: a new tool to improve supply chain efficiency and achieve best practice. Inform Syst Front 3:471–476

Zhou H, Benton WC Jr, Schilling DA, Milligan GW (2011) Supply chain integration and the SCOR model. J Bus Logist 32(4):332–344

Chapter 8
Mathematical Programming Approaches

8.1 Introduction

Mathematical programming is one of the most important techniques available for quantitative decision-making. The general purpose of mathematical programming is finding an optimal solution for allocation of limited resources to perform competing activities. The optimality is defined with respect to important performance evaluation criteria, such as cost, time, and profit. Mathematical programming uses a compact mathematical model for describing the problem of concern. The solution is searched among all feasible alternatives. The search is executed in an intelligent manner, allowing the evaluation of problems with a large number of feasible solutions.

Mathematical programming finds many applications in supply chain management, at all decision-making levels. It is also widely used for supply chain configuration purposes. Out of several classes of mathematical programming models, mixed-integer programming models are used most frequently. Other types of models, such as stochastic and multi-objective programming models, are also emerging to handle more complex supply chain configuration problems. Although these models are often more appropriate, computational complexity remains an important issue in the application of mathematical programming models for supply chain configuration.

This chapter describes application of mathematical programming for supply chain configuration. The general overview is given in Sect. 8.2. It is followed by a description of generic supply chain configuration mixed-integer programming model in Sect. 8.3. This model is based on the data model presented in Chap. 7. Computational approaches for solving problems of large size are also discussed along with typical modifications of the generic model, especially, concerning global factors. Section 8.4 outlines the application of other classes of mathematical programming models. In Sect. 8.5, the generic optimization model is used to optimize the SCC Bike's supply chain configuration. Section 8.6 details a model

© Springer Science+Business Media New York 2016
C. Chandra, J. Grabis, *Supply Chain Configuration*,
DOI 10.1007/978-1-4939-3557-4_8

integration procedure, whereby optimization models for supply chain configuration problems can be built on the basis of pertinent information models.

8.2 Purpose

Mathematical programming models are used to optimize decisions concerning execution of certain activities subject to resource constraints. Mathematical programming models have a well-defined structure. They consist of mathematical expressions representing objective function and constraints. The expressions involve parameters and decision variables. The parameters are input data, while the decision variables represent the optimization outcome. The objective function represents modeling objectives and makes some decisions more preferable than others. The constraints limit the values that decision variables can assume.

The main advantages of mathematical programming models are that they provide a relatively simple and compact approximation of complex decision-making problems, an ability to efficiently find an optimal set of decisions among a large number of alternatives, and supporting analysis of decisions made. Specifically, in the supply chain configuration problem context, mathematical programming models are excellent for modeling its spatial aspects.

There are also some important limitations. Mathematical programming models have a lower level of validity compared to some other types of models—particularly, simulation. In the supply chain configuration context, mathematical programming models have difficulties representing the dynamic and stochastic aspects of the problem. Additionally, solving of many supply chain configuration problems is computationally challenging.

Following the supply chain configuration scope, mathematical programming models are suited to answer the following supply chain configuration questions:

1. Which partners to choose?
2. Where to locate supply chain facilities?
3. How to allocate production and capacity?
4. Which transportation mode to choose?
5. How do specific parameters influence supply chain performance?

The most common type of mathematical programming models is linear programming models. These models have all constraints and the objective function expressed as a linear function in variables. However, many real-life problems cannot be represented as linear functions. A typical example is representation of decisions concerning the opening of supply chain facilities. These decisions assume values equal either to 0 or 1. Integer programming models are used to model such problems. Their computational tractability is lower than that of linear programming models. Nonlinear expressions are often required to represent inventory and transportation-related issues of supply chain. That results in nonlinear programming models, which have high computational complexity.

Given the heterogeneous nature of supply chains, optimization often cannot be performed with respect to a single objective. Multi-objective programming models seek an optimal solution with regard to multiple objectives. These models rely on judgmental assessment of the relative importance of each objective.

Generally, as one moves from linear programming to more complex mathematical programming models, the validity of representing real-world problems is improved at the expense of model development and solving simplicity. Specialized model-solving algorithms are often required to solve complex problems.

Mathematical programming modeling systems (Greenberg 1993) have been developed for elaboration, solving, and analysis of mathematical programming models. These include GAMS, ILOG, and LINGO, to mention a few. These systems provide the means for data handling, model composition using special-purpose mathematical programming languages, and model solving. From the perspective of integrated decision modeling frameworks, these systems can be easily integrated into the decision support system to provide optimization functionality. The integration is achieved by using some types of application programming interfaces. Data structures used, generally, are system specific. Therefore, these need to be mapped to data sources using information modeling.

The role of mathematical programming systems in the overall strategic decision-making system has been described by Shapiro (2006). The described optimization modeling system includes links from the mathematical programming system to a decision-making database and other data sources, as well as advanced tools for conducting analysis. Generation of optimization models from data stored in the decision-making database is considered.

8.3 Mixed-Integer Programming Models

Traditional supply chain configuration models are mixed integer programming models. This section starts with presenting a generic model formulation which includes only the most frequently used decision variables, parameters, and constraints, as identified during construction of the generic supply chain configuration data model. The presentation of the generic model is followed by an overview of most frequently used modifications.

8.3.1 Generic Formulation

The following subsections define notation used to specify the generic supply chain configuration optimization model, and present the object function and constraints of this model.

Notations

Notation	Definition
Indices	
i	Products, $i = 1, \ldots, I$
j	Materials, $j = 1, \ldots, J$
k	Plants, $k = 1, \ldots, K$
s	Suppliers, $s = 1, \ldots, S$
m	Distribution centers, $m = 1, \ldots, M$
n	Customer zones, $n = 1, \ldots, N$
t	Time period, $t = 1, \ldots, T$
Parameters	
d_{int}	Demand
π_i	Revenues per product
h_k	Plant capacity
γ_i	Capacity requirements for product
δ_{ij}	Material consumption per product
ρ_{js}	$\rho_{js} = 1$ if the supplier offers a material and $\rho_{js} = 0$ otherwise
ω_{js}	Material purchasing cost from supplier per unit
λ_{ik}	Production cost at plant per unit
β_{im}	Inventory storage cost at the distribution center over the planning horizon per product
r_{im}	Handling cost at distribution center per unit
c_{1jsk}	Transportation cost from supplier to plant per material unit
c_{2ikm}	Transportation cost from plant to distribution center per product unit
c_{3imn}	Transportation cost from distribution center to customer per product unit
f_{1k}	Plant fixed opening/operating cost per time period
f_{2m}	Distribution center fixed opening/operating cost per time period
P	A large constant number
Decision variables	
X_{imnt}	Quantity of products sold from distribution center to customer
Q_{ikt}	Quantity of products produced at plant
Y_{ikmt}	Quantity of products shipped from plant to distribution center
B_{imt}	Inventory size at distribution center per product
V_{jskt}	Quantity of materials purchased and shipped from supplier to plant
W_k	Plant open indicator equals 1 if plant is open and 0 otherwise
U_m	Distribution center open indicator equals 1 if distribution center is open and 0 otherwise
A_{mn}	Customer zone to distribution center allocation indicator equals 1 if customer is served by distribution center and 0 otherwise

Objective Function

The objective function (Eq. 8.1) maximizes profit E determined as a difference between revenues Φ and total cost TC. As indicated in the previous chapter, profit maximization increasingly is considered as one of the main supply chain configuration performance measures. The total cost consists of multiple cost components including production cost (TC_1), materials purchasing and transportation cost (TC_2), products transportation cost from plants to distribution centers, product handling and transportation cost from distribution centers to customers (TC_3), fixed costs for opening and operating plants and distribution centers (TC_4), and inventory holding cost (TC_5). Revenues, total cost, and its components collectively are referred as measures used to evaluate supply chain configuration performance.

$$E = \Phi - TC \to \max \tag{8.1}$$

Measures

$$\Phi = \sum_{i=1}^{I} \sum_{m=1}^{M} \sum_{n=1}^{N} \sum_{t=1}^{T} \pi_i X_{imnt} \tag{8.2}$$

$$TC = \sum_{l=1}^{5} TC_l \tag{8.3}$$

$$TC_1 = \sum_{i=1}^{I} \sum_{k=1}^{K} \sum_{t=1}^{T} \lambda_{ik} Q_{ikt} \tag{8.4}$$

$$TC_2 = \sum_{j=1}^{J} \sum_{s=1}^{S} \sum_{k=1}^{K} \sum_{t=1}^{T} \left(\omega_{js} + c_{1jsk} \right) V_{jskt} \tag{8.5}$$

$$TC_3 = \sum_{i=1}^{I} \sum_{k=1}^{K} \sum_{m=1}^{M} \sum_{t=1}^{T} c_{2ikm} Y_{ikmt}$$
$$+ \sum_{i=1}^{I} \sum_{m=1}^{M} \sum_{n=1}^{N} \sum_{t=1}^{T} \left(r_{im} + c_{3imn} \right) X_{imnt} \tag{8.6}$$

$$TC_4 = \sum_{k=1}^{K} f_{1k} W_k + \sum_{m=1}^{M} f_{2m} U_m \tag{8.7}$$

$$TC_5 = T^{-1} \sum_{i=1}^{I} \sum_{m=1}^{M} \sum_{t=1}^{T} \beta_{im} B_{imt} \tag{8.8}$$

Constraints

$$\sum_{m=1}^{M} X_{imnt} \le d_{int}, \forall i, n, t \tag{8.9}$$

$$\sum_{n=1}^{N} X_{inmt} \le B_{imt} + \sum_{k=1}^{K} Y_{ikmt}, \forall i, m, t \tag{8.10}$$

$$B_{imt} = B_{imt-1} + \sum_{k=1}^{K} Y_{ikmt} - \sum_{n=1}^{N} X_{inmt}, \forall i, m, t \tag{8.11}$$

$$\sum_{m=1}^{M} Y_{ikmt} \le Q_{ikt}, \forall i, k, t \tag{8.12}$$

$$\sum_{i=1}^{I} \gamma_i Q_{ikt} \le h_k W_k, \forall k, t \tag{8.13}$$

$$\sum_{i=1}^{I} \delta_{ij} Q_{ikt} \le \sum_{s=1}^{S} \rho_{js} V_{jskt}, \forall j, k, t \tag{8.14}$$

$$\sum_{i=1}^{I} \sum_{n=1}^{N} X_{imnt} < PU_m, \forall m, t \tag{8.15}$$

$$\sum_{n=1}^{N} A_{mn} = 1, \forall n \tag{8.16}$$

$$B_{im0} = 0, \forall i, m \tag{8.17}$$

$$W_k, U_m \in \{0, 1\}, \forall k, m \tag{8.18}$$

$$A_{mn} \in \{0, 1\}, \forall m, n \tag{8.19}$$

Equation (8.9) enforces the balance between products sold and demand. The balance between incoming and outgoing flows at distribution centers is defined by Eqs. (8.10) and (8.11). This balance is achieved by satisfying the customer demand with newly arrived shipments from plants or from the inventory. If some of the newly arrived shipments are not sold to customers, they are retained in inventory at distribution centers. The balance between products produced and products shipped to the distribution centers is enforced by Eq. (8.12). Equation (8.13) restricts capacity availability. Availability of materials to produce products is checked by Eq. (8.14). Equation (8.15) states that product flows are allowed only through open distribution centers. Equation (8.16) control allocation of customer zones to distribution centers by requiring that each customer zone is served by only one distribution center. The initial inventory of products at distribution centers is set to zero Eq. 8.17. Variables W_k, U_m and A_{mn} are binary Eqs. (8.18) and (8.19).

Comments
The model does not explicitly include parameters characterizing a spatial location of supply chain units. Alternative locations for a particular supply chain unit are evaluated by allowing for several units with equal characteristics but different transportation costs, which characterize the location of the unit.

There are two factors affecting the model composition: (1) the broker and power structure; and (2) the initial state of the network. Depending upon the organizational and power structure of the supply chain and a decision maker's point of view (i.e., interests of the whole supply chain vs. interests of the dominant member), some of the cost parameters are set to zero because the total cost the broker is concerned about is not affected by these cost parameters, even if these are relevant to the overall supply chain modeling (e.g., a final assembler pays only purchasing costs for components and is not concerned about processing costs at the supply level). The initial state of the network determines whether some of the decision variables already do not have a fixed value. For instance, the location of several assembly plants is already fixed and cannot be changed. Similarly, long-term purchasing contracts with some suppliers can set definite limits on purchasing volume from these suppliers.

Reconfiguration
The model implicitly assumes that greenfield supply chain configuration is performed and there are no fixed supply chain units or links. In the case of supply chain reconfiguration, additional constraints are imposed to represent the reconfiguration options. If a unit or link is indicated as design time selection, then constraints (8.13) and (8.15)–(8.16) are not changed. If a unit or link is indicated as fixed, then the corresponding constraints are set equal to one (i.e., the decision variable becomes a parameter).

If configuration decision variables n are made at execution time, then the selection variables can assume values either 1 or 0. In this case, it is suggested that design time evaluation of the impact of execution time decisions should be performed by means of robust optimization.

8.3.2 Modifications

The generic formulation obviously needs to be adjusted to include factors relevant to a particular decision-making problem. The literature analysis suggests that the most frequently considered factors are international factors, inventory, capacity treatment, transportation, and supply chain management policies. We discuss these below.

International Factors
Given that many supply chains involve partners from different countries, international factors need to be addressed in supply chain configuration. This problem is of particular importance for large multinational companies manufacturing and selling their products worldwide. Mathematical programming models consider quantitative factors, while there are also numerous qualitative factors influencing international decision-making.

Table 8.1 lists selected decision variables, parameters, and constraints used in some international supply chain configuration models. Goetschalckx et al. (2002) provide a summary table on works considering international factors. This summary indicates that taxes and duties are the most often considered international factors. In a similar work by Meixell and Gargeya (2005), the most frequently considered international factors besides tariffs and duties are currency exchange rates and corporate income taxes. However, many of the models surveyed use already fixed supply chain configuration. Kouvelis et al. (2004) present an extensive sensitivity analysis of the impact of international factors on supply chain configuration. The transfer pricing to optimize overall global supply chain profitability is analyzed in a recent contribution by De Matta and Miller (2015).

Inventory
There has been a significant increase of supply chain configuration models including inventory management related issues as supply chain configuration problem solving. The literature review shows that 26 out of the 68 survey mathematical programming models are multi-period models including inventory management decisions. This trend is driven by an increasing need to analyze supply responsiveness. The inventory management decisions are represented not only at the tactical level but also at the operational decision including safety stock and ordering quantity.

Capacity Treatment
A majority of models have some sort of flow intensity and transformation capacity limits as a parameter. A parameter characterizing capacity consumption per unit processed or handled is also widely used (e.g., Pirkul and Jayaraman 1998;

Table 8.1 Selected international factors considered in literature

Source	International factor
Arntzen et al. (1995)	Duty charge for shipping product on a link
	Tax on product at facility
	Duty drawback credit for a product imported into a nation-group from another nation-group and re-exported in the same condition/different condition
Bhutta et al. (2003)	Exchange rate
	Tariff rate for a product from a facility to market
De Matta and Miller (2015)	Profit tax rate Exchange rate Transfer pricing
ElMaraghy and Mahmoudi (2009)	Currency exchange rate Labor cost
Liu and Papageorgiou (2013)	Import duty rate
Kouvelis et al. (2004)	Income tax rate
	Depreciation rate
	Discount rate of after-tax cash flows

Sabri and Beamon 2000). Sabri and Beamon (2000) and Yan et al. (2003) use product specific capacity, while Pirkul and Jayaraman (1998) the flexible capacity. Bhutta et al. (2003) is one of the few papers using capacity as a decision variable. This paper allows either increasing or decreasing capacity at the facility.

In order to account for environmental factors, it is also important to consider a kind of capacity or resources used in supply chain processes. For instance, Chaabane et al. (2012) set capacity limits for specific production technologies and the most appropriate production technology is used to minimize emissions associated with production as one of the supply chain configuration objectives.

Transportation

The most common way of representing transportation is considering just one mode and including variable costs per unit shipped between supply chain units. However, transportation-related issues generally are much more complex and several models attempt to account for this complexity. Nonlinear dependence of transportation costs according to quantity shipped is modeled by Tsiakis et al. (2001). This dependence is represented by a piece-wise linear function. Transportation costs are not calculated for individual products but for families of similar products, thus reducing the model complexity. Syam (2002) and Viswanadham and Gaonkar (2003) include a fixed charge per unit using a particular link to transfer products between units. Arntzen et al. (1995), Dogan and Goetschalckx (1999), and Viswanadham and Gaonkar (2003) also include the transportation time parameter. Prakash et al. (2012) consider different transportation modes that allows for multi-objective evaluation of the supply chain configuration in order to minimize costs and maximize demand fill rate.

Ross et al. (1998) have transportation as one of the key specific problems of supply chain configuration decision-making and the model represents individual vehicles with their characteristics. Farahani et al. (2015) combine distribution network design with vehicle routing by assigning retailers to a specific delivery route. This approach is particularly useful in the case of agile supply chain, where distribution network design decision are revised relatively often. Vidal and Goetschalckx (2001) split transportation costs between supplier and manufacturer to take advantage of lower taxes.

Capacity limits are also frequently used for links between units. Arntzen et al. (1995) and Syam (2002) represent transportation capacity by limiting the total weight of products shipped. The shipment weight-based representation of shipments costs and transportation capacity is often used in applied studies.

Detailed representation of transportation is a feature of many commercial supply chain network design models. These are based on detailed databases of distance and freight rates. These data as well as transportation cost structure and shipment planning are described by Bowersox et al. (2002).

Supply Chain Management Policies
Configuration decisions concerning use of particular supply chain facilities are often tightly interrelated with strategic-level decisions in relation to the particular managerial policies used. Two cases of representing management policies are distinguished:

- Policies are represented structurally;
- Policies are represented through values of parameters.

An example of structurally represented policies is a decision between using direct shipments and using a centralized warehouse. Evaluation of such alternatives effectively implies development of two separate models, which share common features. However, it is also possible to construct a single model with binary variables used for switching between different structures.

An example of policies represented through values of parameters is a decision between using Electronic Data Interchange (EDI) or the Internet as a communication mode among supply chain units. In this case, a binary variable can be used to represent the decision between policies, and values of parameters representing fixed costs for establishing links among units and variable costs for transferring products are specified for each of the two policies.

A combined example, where policies are represented both structurally and through values of parameters, is a decision variable between using flexible manufacturing facilities or specialized manufacturing facilities. Structurally different product-to-facility assignments are given as inputs (i.e., multiple flexibility scenarios are evaluated). At the same time, representing flexible manufacturing facilities influences the value of the fixed cost parameter.

The literature on including policy-related variables in the quantitative supply chain configuration models is scarce. Truong and Azadivar (2005) include a decision variable representing a choice between using push and pull manufacturing policies.

Analyzing many different policies might lead to explosive growth of the computational time needed to solve the model. Therefore, many policy related decisions are already made at earlier steps of the supply chain configuration.

8.3.3 Computational Issues

Model solving is an important part of supply chain configuration problem solving because the direct use of commercially available solvers might not be sufficient. Geoffrion and Powers (1995), in their discussion of developments in design of integrated production–distribution networks, indicate that corresponding large-scale models are difficult to solve in reasonable time because it is an NP-hard problem. Small to medium problems can be solved using standard software on personal computers (Kouvelis et al. 2004). However, that depends on the structure of a particular model and values of parameters. Specialized model-solving algorithms are generally required to solve large-scale problems.

There are two major approaches to elaboration of computationally efficient algorithms. These are based on Lagrangian relaxation and Bender's decomposition. A short overview of these methods is provided here. Readers are referred to Avriel and Golany (1996) for a detailed coverage of mathematical programming.

Lagrangian Relaxation
The Lagrangian relaxation schema assumes that problem solving is complicated by a few *difficult* constraints. It attempts to simplify the problem by dualizing the difficult constraints (i.e., constraints are introduced into the objective function with a penalty function). As a result, a relaxed problem of the original problem is obtained. The relaxed problem is solved to obtain an upper bound (for maximization problems) of the original problem. Any feasible solution of the original problem provides a lower bound. Iterative heuristic algorithms are used in searching for the optimal solution of the original problem in this narrowed range. The upper and lower bounds are continuously updated. A good overview of the general theory on the Lagrangian relaxation is provided by Magee and Glover (1996).

Pirkul and Jayaraman (1998) successfully applied the Lagrangian relaxation problem for the supply chain configuration problem. Similar results have been obtained by Jang et al. (2002) and Amiri (2006). The supply chain configuration model by Pirkul and Jayaraman (1998) locates a specified number of manufacturing facilities and warehouses to minimize fixed and transformation costs subject to customer demand satisfaction and capacity constraints.

The mathematical representation of their model is as follows. Parameters of the model are:

C_{ijl}—the variable cost to distribute a unit of product l from warehouse j to customer zone i;

T_{jkl}—a unit cost to transport product l from plant k to warehouse j;

f_k and g_j—fixed cost to open and operate plant k and warehouse j, respectively;
a_{il}—demand for product l at customer zone i;
D_k—capacity of plant k;
W_j—throughput limit at warehouse l;
q_l—plant capacity consumption by product l;
s_l—is warehouse throughput capacity consumption by product l;
W and P—upper limit on the number of warehouses and plants that can be opened, respectively.

Variables X_{ijl} and Y_{jkl} denote the total number of units of product l distributed through warehouse j to customer zone i and the total number of units of product l shipped from plant k to warehouse j, respectively. P_k and Z_j are binary variables denoting whether plant k is open and whether warehouse j is open, respectively.

The objective function and constraints are given below.

$$\min Z = \sum_i \sum_j \sum_l C_{ijl} X_{ijl} + \sum_j \sum_k \sum_l T_{jkl} X_{jkl} + \sum_k f_k P_k + \sum_j g_j Z_j \quad (8.20)$$

subject to

$$\sum_j X_{ijl} = a_{il}, \forall i, l \quad (8.21)$$

$$\sum_i \sum_l s_l X_{ijl} \leq Z_j W_j, \forall j \quad (8.22)$$

$$\sum_j Z_j \leq W \quad (8.23)$$

$$\sum_i X_{ijl} \leq \sum_k Y_{jkl}, \forall j, l \quad (8.24)$$

$$\sum_i \sum q_l Y_{jkl} \leq D_k P_k, \forall k \quad (8.25)$$

$$\sum_k P_k \leq P, \forall k \quad (8.26)$$

After relaxing constraints Eqs. (8.21) and (8.24), the Lagrangian relaxation of the problem is

$$\min Z_{LR} = \sum_i \sum_j \sum_l C_{ijl} X_{ijl} + \sum_j \sum_k \sum_l T_{jkl} X_{jkl} + \sum_k f_k P_k + \sum_j g_j Z_j$$
$$+ \sum_i \sum_l \gamma_{il} \left(\sum_j X_{ijl} - a_{il} \right) + \sum_j \sum_l \beta_{jl} \left(\sum_i X_{ijl} - \sum_k Y_{jkl} \right)$$
$$(8.27)$$

where γ_{il} and β_{jl} are Lagrangian multipliers (dual prices). The relaxed problem is further decomposed into a subproblem representing manufacturing plants and a subproblem representing warehouses. An iterative model-solving procedure is used to solve the configuration problem. The Lagrangian subproblems are used to narrow the gap between lower and upper bounds until the difference is less than one percent or 500 iterations have been executed. Computational efficiency of the procedure has been tested for different numbers of products, potential plants and warehouses, and customer zones, as well as for different levels of capacity load. For instance, the problem-solving time for a problem with 100 customer zones, 20 warehouses, 10 plants and 3 products is about 60 s.

Bender's Decomposition
The main idea behind the Bender's decomposition approach is partitioning the original mixed-integer problem into its linear and integer parts (Salkin 1975). The steps of the problem-solving algorithm are as follows:

1. Fix values of integer variables and determine upper and lower bounds.
2. Solve a dual problem of the linear programming model obtained by fixing the integer variables and update the upper bound (for minimization problems).
3. Solve an integer problem obtained from the original problem by fixing the continuous part of the problem and update lower bound.
4. Iterate until the gap between the upper and lower bound is sufficiently small.
5. Upon convergence, compute optimal values of continuous decision variables.

At the first step, not only integer variables can be fixed but also any variables deemed as complicated. The Benders decomposition for solving supply chain configuration problems has been used by Geoffrion and Graves (1974), and Dogan and Goetschalckx (1999). In both cases, it has allowed solving large industrial-scale problems within a reasonable time. The former authors additionally develop a specialized acceleration technique, which has been shown to decrease computational time substantially.

8.4   Other Mathematical Programming Models

Multi-objective, stochastic, and nonlinear mathematical programming models are other models that find application in supply chain configuration.

8.4.1   Multi-objective Programming Models

A multi-objective evaluation is needed to represent various aspects of supply chain performance and customers' requirements satisfaction, as well as to balance the performance of individual supply chain units. Two main technical approaches to representing multi-objective situations are: (1) assigning weights to each objective, characterizing relative importance; and (2) preemptive optimization starting with the most important objective. Choice of appropriate weights and prioritization of objective relies on the decision maker's judgment and substantially affects modeling results.

The generic formulation can be extended to multi-objective setting in various ways. Objectives associated with environmental issues, responsiveness, and reliability including customer service are considered most frequently. The generic supply chain configuration optimization model is extended to incorporate these additional objectives and the objective function (8.1) is reformulated as

$$Z = \min(Z_1, Z_2, Z_3, Z_4), \tag{8.28}$$

where $Z_1 = TC$ represents total costs, Z_2 represents environmental impact, Z_3 represents responsiveness and Z_4 represents reliability (other notation is used as in Sect. 8.3.1 but omitting time period index t). The environmental impact is evaluated as quantity of carbon emissions due to transportation

$$Z_2 = \sum_{j=1}^{J} \sum_{s=1}^{S} \sum_{k=1}^{K} e_j \tau_{1jsk} V_{jsk}$$
$$+ \sum_{i=1}^{I} \sum_{k=1}^{K} \sum_{m=1}^{M} e_i \tau_{2ikm} Y_{ikm} + \sum_{i=1}^{I} \sum_{m=1}^{M} \sum_{n=1}^{N} e_i \tau_{3imn} X_{imn} \to \min, \tag{8.29}$$

where e_j and e_i are carbon emissions associated with transportation of materials and products, respectively, and τ_{1jsk} is transportation time from supplier to plant per material unit, τ_{2ikm} is transportation time from plant to distribution center per product unit, and τ_{3mn} is transportation time from distribution center to customer per product unit.

The responsiveness is evaluated as a time spent during transportation of materials and products along the supply chain links

$$Z_3 = \sum_{j=1}^{J} \sum_{s=1}^{S} \sum_{k=1}^{K} t_{1jsk} V_{jsk}$$
$$+ \sum_{i=1}^{I} \sum_{k=1}^{K} \sum_{m=1}^{M} t_{2ikm} Y_{ikm} + \sum_{i=1}^{I} \sum_{m=1}^{M} \sum_{n=1}^{N} t_{3imn} X_{imn} \to \min \tag{8.30}$$

The supply chain reliability is evaluated by the fill rate

$$Z_4 = D^{-1} \sum_{i=1}^{I} \sum_{m=1}^{M} \sum_{n=1}^{N} X_{imn} \to \max \tag{8.31}$$

where $D = \sum_{i=1}^{I} \sum_{n=1}^{N} d_{in}$ is the total demand for all products in the supply chain.

Other multi-objective supply chain configuration models have been developed by Li and O'Brien (1999), Sabri and Beamon (2000), Talluri and Baker (2002), Brandenburg (2015) and Das and Rao Posinasetti (2015) (see Chap. 3).

8.4.2   Stochastic Programming Models

The models discussed above assume that all parameters are known with certainty, which is not the case in real-life situations. To obtain robust results, the impact of uncertainty needs to be assessed. Stochastic programming is one of the techniques allowing accounting for stochastic parameters.

Many of the stochastic programming models developed for supply chain configuration have demand as a stochastic parameter. Demand uncertainty usually is represented by multiple demand scenarios (Mirhassani et al. 2000; Tsiakis et al. 2001). In this case, a prototype objective function can be expressed as

$$Z = \max_{\mathbf{Q},\mathbf{Y}} E[F(\mathbf{c},\mathbf{D},\mathbf{Q},\mathbf{Y})] = \max_{\mathbf{Q},\mathbf{Y}} \sum_{s=1}^{S} F(\mathbf{c},\mathbf{D}_s,\mathbf{Q},\mathbf{Y}), \qquad (8.32)$$

where $\mathbf{c}$ represents all parameters of the supply chain configuration problem, $\mathbf{D}$ represents demand, $\mathbf{Q}$ represents continuous decision variables and $\mathbf{Y}$ represents binary decision variables (e.g., inclusion of units in the supply chain). F is an abstract function, E is the expected profit, and $s = 1,\ldots,S$ are evaluated demand scenarios.

Other stochastic parameters can also be represented by evaluation of multiple scenarios (e.g., Gutiérrez et al. 1996). The obvious limitation of this approach is a limited number of considered scenarios and there is little assurance that the coverage of uncertainty has been adequate.

Kim et al. (2002) develop a model for determining ordering quantities from suppliers for a fixed supply chain network subject to demand uncertainty. The demand uncertainty is represented using demand probability density function and an iterative model-solving procedure is developed without relying on using scenarios.

Santoso et al. (2005) develop a stochastic programming model for a typical supply chain configuration problem. The model minimizes total investment and operating costs by deciding which facilities to build and routing products from suppliers to customers. It allows for uncertainty in processing/transportation costs, demand, supplies, and capacities and for limited, but a very large number of scenarios representing uncertainty in demand, as well as in other parameters. The main constraints enforce capacity limits, flow conversion limits, and facility opening requirements (i.e., facility is operational only if open). The model is a two-stage stochastic program that minimizes the current investment cost and expected operational costs.

A specialized model-solving algorithm is developed. It uses an accelerated Benders decomposition to solve the facility opening problem and the sample average approximation scheme to solve the stochastic part of the model. The model is tested by its application in designing a supply chain in the packaging industry. The authors show that the developed model-solving algorithm allows solving large scale problems (13 products and 142 facilities) in less than two hours for one scenario and, more importantly, growth of computational time as the number of scenarios increases is slow. The stochastic approach to supply chain design allowed savings of up to 6 % compared to the mean value problem solution for the considered supply chain design problem. The stochastic programming solution also exhibits substantially lower variability over testing scenarios, which is a desirable property during the results approbation phase of the supply chain configuration methodology.

8.4.3 Nonlinear Programming Models

Due to major computational difficulties, nonlinear configuration models have not been frequently encountered in the supply chain configuration literature (see Wu and O'Grady (2004) for a brief discussion of nonlinear programming models in supply chain configuration). The main nonlinear factors relevant to supply chain configuration, such as inventory and transportation costs, are usually represented using piece-wise linear functions (e.g., Tsiakis et al. 2001).

Explicitly, nonlinear constraints have been used in models solved using simulation-based optimization and other nonparametric optimization methods, which are discussed in Chap. 9.

8.5 Sample Application

To illustrate application of the generic mixed-integer programming model presented in Sect. 8.3.1, the SCC Bike supply chain configuration is optimized. The objective is to maximize profit by selecting suppliers, locating the assembly plants and allocating customers to distribution centers. Two product groups are considered during the configuration and they differ mainly by the type of frame used in their production. The quarterly demand exhibits seasonal variations (Table 8.2) and manufacturing capacity is not sufficient to handle demand peeks in a single period (Table 8.3). The revenues per product also account for costs not explicitly considered in the configuration model. The production cost is higher for carbon frame bikes and is randomized to induce differences among the productions sites. The fixed cost is determined according to industry data concerning invest-ments made per plant of certain capacity[1] and adjusted to include operational costs.

There are at least two suppliers for every set of materials. The suppliers vary by prices offered (generated by randomly around a specified mean value) and by location what affects the transportation cost. The transportation cost for materials is generated by assuming that a sea transport is used at the cost of $0.03 per

---

[1] http://www.bicycleretailer.com/.

Table 8.2 Quarterly demand for product groups

Product group	π_i ($)	d_{i1}	d_{i2}	d_{i3}	d_{i4}
Alumina frame bikes	880	25,000	100,000	50,000	25,000
Carbon frame bikes	1350	6250	25,000	12,500	6250

Table 8.3 Manufacturing capacity and production costs at plants

	Manufacturing capacity (units/period)	Production cost ($/units)	Fixed cost ($/period)
Frame factory	100,000	384–673	800,000
Assembly plant (US)	60,000	220–250	1,500,000
Assembly plant (EU)	25,000	250–375	800,000

Table 8.4 Material prices offered by different suppliers ($/set)

Materials	MCS1	OIS1	NMPS1	MCS2	MCS3	NMPS2	OIS2	NMPS3
Mechanical components parts	161	0	0	167	170	0	0	0
Other industries parts	0	56	0	0	0	0	43	0
Non-moving parts	0	0	106	0	0	132	0	139

thousand units per mile (twice as much for frames). The actual sea-link distances are used to calculate c_1. The material purchasing prices are given in Table 8.4. Transportation costs for products are generated assuming that trucks are used for transportation at the cost of $2.8 and $5.6 per thousand units per mile. These transportation cost coefficients and actual distances between locations are used to calculate c_2 and c_3, respectively.

The optimization is performed for the base scenario characterized by the data provided above. The optimized profit $E = 74.4$ mil.$ and the service level measured as a ratio between total sales and total demand is 90 %. The service level is below 100 % due to insufficient capacity to deal with seasonal variations in demand. Figure 8.1 shows the costs breakdown according to the measures used. The sourcing costs (purchasing plus material transportation) are the biggest expense and the transportation costs have relatively little impact on the total cost. Therefore, material prices and production costs have the most significant impact on configuration results. The inventory cost is negligible because only period-to-period inventory storage cost at the distribution centers is taken into account (work-in-process inventory and inter-period inventory are not explicitly accounted for). The resulting supply chain configuration is shown in Fig. 8.2. Only one supplier for each set of materials is selected and the purchasing cost is dominates the selection. The plants stock the distribution centers regardless of their location in order to deal with the capacity limitations.

Fig. 8.1 The supply chain configuration costs breakdown

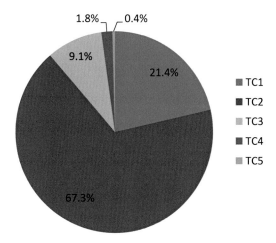

If the capacity restrictions are relaxed or sufficiently high plant to distribution center allocation fee is introduced, the plants would supply products only to their regional distribution centers. The current arrangement of flexible allocation requires distribution centers to cope with regional differences due to different assembly locations (i.e., multi-language user manuals). The seasonal character of the demand and incorporating of inventory management decisions substantially affects the configuration decisions.

8.6 Model Integration

The supply chain configuration methodology emphasizes the integration of decision-making models with information models. Therefore, the supply chain configuration model's data model is used to develop the supply chain optimization model. Figure 8.3 elaborates the transition from information modeling to quantitative modeling. This figure represents implementation of the optimization-related functionality of the integrated decision support system presented in Chap. 5. The commercially available LINGO[2] mathematical programming system is used in this case, although the approach is similar to several other mathematical programming languages. The figure shows only one-way interactions for simplicity. Obviously, modeling outcomes can be sent back to the supply chain management information system in a similar manner.

The general data model discussed previously is developed using a general modeling method such as UML, while the mathematical model is implemented using a special-purpose modeling language, LINGO. The LINGO model includes data definitions in the form of data sets, data link definitions providing link to data

---

[2] www.lindo.com.

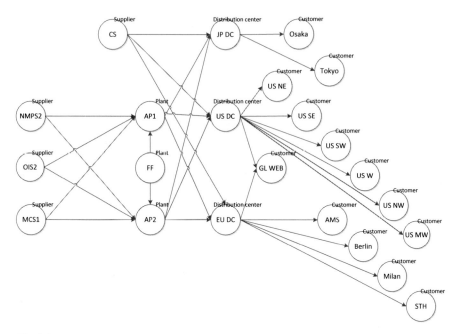

Fig. 8.2 The optimized supply chain configuration

sources, and a formalized representation of the mathematical program. Data definitions and data links are generated automatically using data provided in the general
data model (Fig. 8.4). Transformations are informally listed as follows (numbers in
the list correspond to the numbering of arrows in Fig. 8.4):

1. A data set declaration instruction is generated for each class in the diagram. All
   attributes except *dimension* are also included in the instruction line to declare
   parameters and variables of the mathematical programming model.
2. A variable declaration instruction is generated for the *dimension* attribute of
   each class (in the example, the variable is named *PD*). The generated instruction
   also defines data reading from the data source (using the @POINTER function
   of the special purpose programming language).
3. An instruction for reading values of the declared parameters is generated for
   each attribute of the *Parameter* type in the class diagram.
4. Attributes of type *DecisionVariable* are only included in the data set declaration
   instruction line (see Transformation 1 above) and this arrow only signifies the
   representation of decision variables.

The mathematical program is composed in a semiautomated manner by a
decision maker who indicates which constraints to include from the decision-
modeling knowledge base. The modeling technique specific data model contains
actual data to be passed from the decision-modeling system to the LINGO solver
during the problem-solving process. LINGO supports two main data transfer
mechanisms:

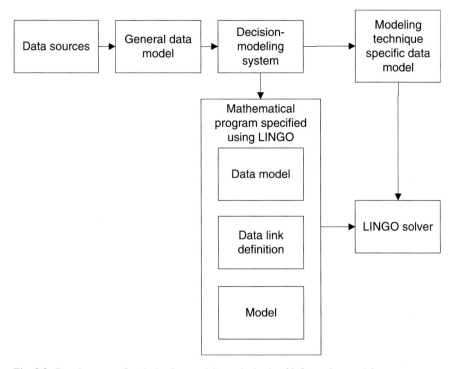

Fig. 8.3 Development of optimization models on the basis of information models

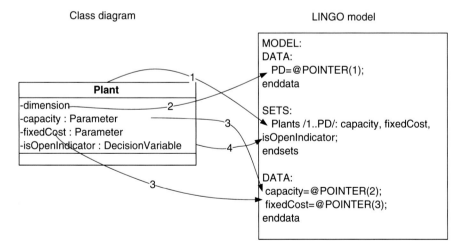

Fig. 8.4 Generation of the LINGO data definition from the general data model

- *Open Database Connectivity (ODBC) based data transfer.* In this case, the separate modeling technique specific data model is not necessary because LINGO can directly request data from database tables using the standard database access protocol.
- *Remote Procedure Call (RPC) based data transfer.* This mode is necessary if LINGO is part of a more complex decision-making system and is invoked programmatically. In this case, LINGO receives two specially structured data arrays from the decision-modeling system. The first array contains meta-data about data being transferred. The second array contains actual values. The decision-modeling system is responsible for merging data from the general data model into these two arrays.

8.7   Summary

This chapter describes the generic supply chain configuration model, modifications of this model, and the integration of the mathematical programming model into the overall decision-modeling process.

Computational limitations still remain an important factor when considering practical application of mathematical programming for the supply chain configuration problem solving. Solving configuration models using computational approaches described in this chapter requires substantial expertise in mathematical programming, and algorithms are developed on a case-by-case basis. Therefore, commercial applications often rely on pure computational power or heuristic approaches. The former is not always sufficient for medium-size problems, while the latter cannot guarantee the quality of obtained solutions. Computational feasibility also restricts the development of nonlinear mathematical programming models.

References

Amiri A (2006) Designing a distribution network in a supply chain system: formulation and efficient solution procedure. Eur J Oper Res 171:567–576

Arntzen BC, Brown GG, Harrison TP, Trafton LL (1995) Global supply chain management at digital equipment corporation. Interfaces 25:69–93

Avriel M, Golany B (1996) Mathematical programming for industrial engineers. Marcel Dekker, New York

Bhutta KS, Huq F, Frazier G, Mohamed Z (2003) An integrated location, production, distribution and investment model for a multinational corporation. Int J Prod Econ 86:201–216

Bowersox DJ, Closs DJ, Cooper MB (2002) Supply chain logistics management. McGraw-Hill, Boston

Brandenburg M (2015) Low carbon supply chain configuration for a new product – a goal programming approach. Int J Prod Res 53:1–23

Chaabane A, Ramudhin A, Paquet M (2012) Design of sustainable supply chains under the emission trading scheme. Int J Prod Econ 135:37–49

Das K, Rao Posinasetti N (2015) Addressing environmental concerns in closed loop supply chain design and planning. Int J Prod Econ 163:34–47

De Matta R, Miller T (2015) Formation of a strategic manufacturing and distribution network with transfer prices. Eur J Oper Res 241:435–448

Dogan K, Goetschalckx M (1999) A primal decomposition method for the integrated design of multi-period production-distribution systems. IIE Trans 31:1027–1036

ElMaraghy HA, Mahmoudi N (2009) Concurrent design of product modules structure and global supply chain configurations. Int J Comput Integrated Manuf 22:483–493

Farahani RZ, Rashidi Bajgan H, Fahimnia B, Kaviani M (2015) Location-inventory problem in supply chains: a modelling review. Int J Prod Res 53:3769–3788

Geoffrion AM, Graves GW (1974) Multicommodity distribution system design by Benders decomposition. Manag Sci 20:822–844

Geoffrion AM, Powers RF (1995) Twenty years of strategic distribution system design: an evolutionary perspective. Interfaces 25:105–128

Goetschalckx M, Vidal C, Dogan K (2002) Modeling and design of global logistics systems: a review of integrated strategic and tactical models and design algorithms. Eur J Oper Res 143:1–18

Greenberg HJ (1993) Enhancements in of ANALYZE: a computer-assisted analysis system for linear programming. ACM Trans Math Software 19:223–256

Gutiérrez GJ, Kouvelis P, Kurawarwala AA (1996) A robustness approach to uncapacitated network design problems. Eur J Oper Res 94:362–376

Jang Y-J, Jang S-Y, Chang B-M, Park J (2002) A combined model of network design and production/distribution planning for a supply network. Comput Ind Eng 43:263–281

Kim B, Leung JMY, Park KT, Zhang GQ, Lee S (2002) Configuring a manufacturing firm's supply network with multiple suppliers. IIE Trans 34:663–677

Kouvelis P, Rosenblatt MJ, Munson CL (2004) A mathematical programming model for global plant location problems: analysis and insights. IIE Trans 36:127–144

Li D, O'Brien C (1999) Integrated decision modelling of supply chain efficiency. Int J Prod Econ 59:147–157

Liu S, Papageorgiou LG (2013) Multiobjective optimisation of production, distribution and capacity planning of global supply chains in the process industry. Omega 41:369–382

Magee TM, Glover F (1996) Integer programming. In: Avriel M, Golany B (eds) Mathematical programming for industrial engineers. Marcel Dekker, New York, pp 123–270

Meixell MJ, Gargeya VB (2005) Global supply chain design: a literature review and critique. Transport Res E Logist 41:531–550

Mirhassani SA, Lucas C, Mitra G, Messina E, Poojari CA (2000) Computational solution of capacity planning models under uncertainty. Parallel Comput 26:511–538

Pirkul H, Jayaraman V (1998) A multi-commodity, multi-plant, capacitated facility location problem: formulation and efficient heuristic solution. Comput Oper Res 25:869–878

Prakash A, Chan FTS, Liao H, Deshmukh SG (2012) Network optimization in supply chain: a KBGA approach. Decis Support Syst 52:528–538

Ross A, Venkataramanan MA, Ernstberger KW (1998) Reconfiguring the supply network using current performance data. Decis Sci 29:707–728

Sabri EH, Beamon BM (2000) A multi-objective approach to simultaneous strategic and operational planning in supply chain design. Omega 28:581–598

Salkin HM (1975) Title integer programming. Addison-Wesley, Reading

Santoso T, Ahmed S, Goetschalckx M, Shapiro A (2005) A stochastic programming approach for supply chain network design under uncertainty. Eur J Oper Res 167:96–115

Shapiro JF (2006) Modeling the supply chain. Duxbury Press, New York

Syam SS (2002) A model and methodologies for the location problem with logistical components. Comput Oper Res 29:1173–1193

Talluri S, Baker RC (2002) A multi-phase mathematical programming approach for effective supply chain design. Eur J Oper Res 141:544–558

Truong TH, Azadivar F (2005) Optimal design methodologies for configuration of supply chains. Int J Prod Res 43:2217–2236

Tsiakis P, Shah N, Pantelides CC (2001) Design of multi-echelon supply chain networks under demand uncertainty. Ind Eng Chem Res 40:3585–3604

Vidal CJ, Goetschalckx M (2001) Global supply chain model with transfer pricing and transportation cost allocation. Eur J Oper Res 129:134–158

Viswanadham N, Gaonkar RS (2003) Partner selection and synchronized planning in dynamic manufacturing networks. IEEE Trans Robot Autom 19:117–130

Wu T, O'Grady P (2004) A methodology for improving the design of a supply chain. Int J Comput Integrated Manuf 17:281–293

Yan H, Yu Z, Cheng TCE (2003) A strategic model for supply chain design with logical constraints: formulation and solution. Comput Oper Res 30:2135–2155

Chapter 9
Simulation Modeling and Hybrid Approaches

9.1 Introduction

Mathematical programming models are described in Chap. 8 as the primary type of models used in supply chain configuration. However, these models have several limitations. Therefore, the integrated supply chain reconfiguration framework and the supply chain configuration methodology consider simulation modeling as an approach to address decision-making issues not covered by mathematical programming models. It is widely recognized that simulation can describe complex systems in a highly realistic manner and is used to explore the properties of such systems.

Simulation is perceived as an essential part of the supply chain configuration process. It complements findings made using mathematical programming and other modeling approaches. In view of the high costs associated with the implementation of configuration decisions, simulation provides the means for detailed evaluation of these decisions before their physical implementation.

A majority of simulation models of supply chain configuration treat the supply network as fixed and alternative configuration are evaluated using the scenario based approach. That limits the number of alternative to be evaluated. In order to overcome this limitation, simulation models can be combined with optimization models to form a hybrid supply chain configuration model. The hybrid modeling approach capitalizes on the strengths of simulation and optimization methods while avoiding weaknesses of these methods. The hybrid modeling is not limited to combining only optimization and simulation models, which remain the cornerstone of hybrid modeling for supply chain configuration, but also includes other models such as analytical hierarchical process and statistical models. It is also a very computationally intensive problem solving approach. Therefore, methods for accelerating the modeling process are highly desirable.

Section 9.2 discusses the purpose of using simulation and hybrid modeling in supply chain configuration. The development of supply chain configuration discrete event simulation models is discussed in Sect. 9.3. The development approach

© Springer Science+Business Media New York 2016

C. Chandra, J. Grabis, *Supply Chain Configuration*,
DOI 10.1007/978-1-4939-3557-4_9

utilizes the concept of generic supply chain unit as the main building block of the simulation model. Section 9.4 defines the type of hybrid models and elaborates a meta-modeling based method for development of the hybrid models. The method allows for computationally efficient integration between the optimization and simulation models. An example illustrating the hybrid modeling is provided.

9.2 Purpose

Simulation is used along with mathematical programming in the selection step of the supply chain configuration methodology. Generally, it is assumed that mathematical programming is used to establish supply chain configuration while simulation is used to evaluate this configuration. The main reason why simulation is used to evaluate decisions made by mathematical programming models is its ability to represent supply chain in a more realistic manner. Simulation can be perceived as a test-bed for implementing configuration decisions. It enables supply chain evaluation with respect to various factors; particularly those representing supply chain dynamic, and stochastic factors and interactions among supply chain units. Simulation also allows obtaining multiple performance measures characterizing both cost and time related characteristics of the supply chain.

Referring back to the definition of the supply chain configuration scope (Chap. 2), simulation is particularly well suited to address managerial concerns, such as:

- Customer service and delivery reliability.
- Quantification of risk factors.
- What-if analysis.

Some important performance measures provided by simulation are:

- Product cycle time.
- Customer service level.
- Probability distributions of cost and time estimates.
- Supply chain robustness.

The last measure is particularly important in the case of reconfigurable supply chains because it also characterizes processes during transition from one supply chain configuration to another.

Despite the powerful capabilities provided by simulation it is rarely used as a standalone tool for solving configuration problems. This is due to several shortcomings of simulation modeling. The main limitation in the supply chain configuration context is that simulation is primarily a descriptive tool, which requires a human decision-maker to identify alternative configurations that he or she wishes to explore. While it is possible to identify such alternatives in some situations, that is not possible in the general case because the number of alternatives is large. Some other disadvantages of simulation modeling are expensive model development and usage, and interpretation of stochastic modeling results.

Simulation modeling can be performed at various levels of abstraction, which is one of the main options to balance model development cost and usability against model validity. Supply chain configuration is a strategic decision-making problem. Therefore, the level of abstraction generally could be kept quite high. Multiple supply chain simulation approaches discussed later in this chapter attempt to generalize supply chain units representing them as abstract nodes in the simulation model. The level of abstraction is likely to vary according to the broker's perspective because sufficient information is likely to be available only about an organization represented by the broker. In the context of the integrated supply chain configuration framework (see Chap. 4), each potential supply chain partner could maintain its simulation model, which can be linked together, if necessary. However, current experiences in executing such inter-enterprise simulation models are limited (Mertins et al. 2005).

Simulation model development and execution forms a subprocess in the selection step of the supply chain configuration methodology. This subprocess begins with problem formulation, data collection, and definition of a conceptual simulation model. That is accomplished by using data provided by the supply chain configuration information models. The sub-process proceeds with the development of an executable simulation model. The executable model is mainly constructed using either general purpose programming languages or specific simulation modeling languages and software packages. The developed model is validated. Upon successful validation, experimental design is constructed and simulation modeling is performed according to this experimental design.

The validation of simulation models is an important, although difficult, step of the simulation modeling process. In the case of supply chain configuration, it is complicated by lack of historical records and a long feedback loop between implementation and observation of results, which are often obscured by other supply chain management decisions. Law and Kelton (2014) list the following approaches to validation of simulation models:

- Structured walkthrough.
- Expert evaluation.
- Comparison against performance of the existing system.
- Comparison against the existing theory.
- Sensitivity analysis.

Comparison with results obtained using other supply chain configuration models is an additional validation technique. Expert evaluation, comparison with manual computations, comparison of simulation with real-life situation, and pilot implementation of simulation results are named as validation methods used by Van der Vorst et al. (2000) in their supply chain redesign model. However, the pilot implementation was possible only for operational level decisions. Bowersox (1972) uses the stability of output data, sensitivity analysis, and comparison of simulated output to historical data as his model validation methods.

9.2.1   Observations

The above overview of selected works on supply chain management allows us to draw the following conclusions about the common characteristics of supply chain simulation:

- The level of abstraction used varies substantially, although it is agreed that a relatively high level of abstraction is sufficient for strategic decision-making purposes. The level of abstraction should be decreased, if the operational level problems need to be evaluated.
- Model development complexity is a major issue in supply chain simulation.
- In response to the previous point, there are attempts to define generic supply chain units and typical elements to be used in a simulation model.
- Supply chain configuration is considered as a fixed input parameter. Configurations are evaluated under numerous scenarios describing supply chain external and internal characteristics.
- A large number of performance measures are used to evaluate supply chain performance. These include those characterizing customer service, inventory management, and other dynamic aspects. Variability of obtained results is assessed.
- Model development is highly case specific and few attempts exist to build upon methodologies reported in the literature.

A comprehensive evaluation of supply chain configuration decisions requires utilization of multiple models as described in previous chapters. There are two alternatives in employing these models: (1) independent models exchanging input–output data, and (2) fully integrated models where selected functions of one model are implemented by another. Application of independent models and interpretation of their often seemingly contradicting results cause difficulties for decision modelers. Therefore, the area of hybrid modeling that can be perceived as the development of a model consisting of two or more highly integrated models is appealing. Such hybrid models exploit strengths of multiple models to provide a single answer to decision modelers.

Hybrid modeling usually considers a combination of analytical and simulation models. In the case of supply chain configuration, analytical modeling is typically implemented using mathematical programming optimization models. However, in general, hybrid modeling is not limited to combining just two models or combining just mathematical programming and simulation models.

Hybrid mathematical programming simulation models are aimed at combining the strengths of mathematical programming and simulation models and reducing the impact of limitations characteristic of these models. Strengths and limitations of such models are summarized in Table 9.1. Mathematical programming models are generally well suited to dealing with spatial issues, while simulation models are more appropriate for dealing with temporal issues. Although the cost of model

Table 9.1 Summary of strengths and limitations of mathematical programming and simulation models

	Mathematical programming models	Simulation models
Strengths	Evaluation of large number of alternative configurations Exact results Capabilities for efficient analysis	Realistic representation of the problem Accounting for dynamic and stochastic factors Availability of different performance measures
Limitations	Quickly increasing model complexity if dynamic, stochastic, and nonlinear factors are added	Expensive development and usage Interpretation of results

development and usage can be substantial for mathematical programming models, it is generally lower than for similar simulation models.

Similar discussion on differences between mathematical models and simulation models can be found in Nolan and Sovereign (1972) and Shanthikumar and Sargent (1983). Ingalls (1998) points out that there are not only technical differences between these two types of models, but the choice between them has major implications from a managerial perspective. Managers are often concerned with finding good workable solutions rather than pursuing more elusive optimal solutions.

If viewed from the supply chain configuration perspective, differences between mathematical programming and simulation are particularly noticeable. Mathematical programming models can be perceived as providing the base decision-making functionality. Their main advantage is the ability to quickly evaluate a large number of alternative configurations. Simulation models, on the other hand, provide the functionality needed to cover the entire scope of the supply chain configuration, especially dynamic and stochastic factors. Given that these factors are very important for reconfigurable supply chains, hybrid models are particularly appropriate for decision-making in this case. Simulation alone can be rarely applied for establishing supply chain configuration because evaluation of a large number of alternative configurations is not computationally feasible.

The application of combined mathematical programming and simulation models has become relatively widespread. Simchi-Levi et al. (2003) describe using an optimization model to establish supply chain configuration and subsequent utilization of simulation to evaluate the established configuration. The procedure is executed in an iterative manner until a sufficiently robust supply chain configuration is found. However, developing a feedback mechanism between optimization and simulation is a challenging theoretical and practical issue, which has limited exposure in the supply chain configuration context. Such feedback mechanisms have been more successfully developed for some other related supply chain management problems. For instance, a production planning hybrid model (Byrne and Bakir 1999) uses a capacity adjustment coefficient to incorporate simulation results in subsequent optimization runs. A formalized feedback mechanism is essential for the success of hybrid modeling, otherwise it downgrades to model combination.

Obviously, fully integrated models also have their own shortcomings, the most significant of which is more complex model development and lower flexibility in model modification and application.

9.3   Development of Simulation Models

Development of simulation models is a complex process. General simulation modeling methodologies have been developed [see Law and Kelton (2014)]. However, simulation models tend to be rather case specific, thus requiring a major development effort. Therefore, specific modeling templates, methods, and tools for a particular problem domain are useful. In the case of supply chain configuration, development of simulation models can be facilitated by exploring several specific characteristics including (1) a high level of abstraction, (2) representing two main elements, namely, supply chain nodes and arcs connecting nodes, and (3) interactions with other supply chain configuration models.

Commercially available simulation packages have attained a high level of maturity. Therefore, using these packages for development of supply chain configuration models and specific utilities facilitating the development process is advisable, instead of relying on custom tools. The benefits of using Commercial-Of-The-Shelf (COTS) software in the framework of optimization and simulation are also identified by Vamanan et al. (2004).

9.3.1   Approach

The proposed simulation model building approach utilizes two main concepts: (1) separation between data and the model; and (2) a generic representation of supply chain units. The main stages of the model building approach are shown in Fig. 9.1.

A supply chain simulation model is developed using data from the supply chain management information system, and is initially specified using UML. If simulation is used to evaluate supply chain configuration optimization results, then optimization results are an important data source. The decision-modeling system generates the simulation model by transforming information models into a specific simulation modeling language, which is generated on the basis of a predefined template. The template does not contain any simulation objects. It only contains procedures for executing control of the generic functions and data declarations. The procedures have a uniform design. Different procedures can be developed to perform the same activity. Thus, different management policies can be analyzed.

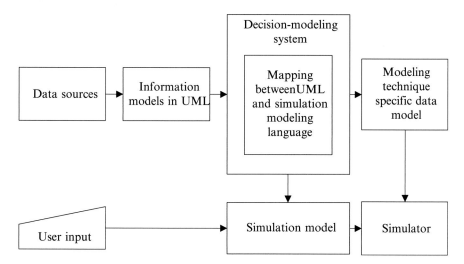

Fig. 9.1 Integrated simulation model building

The generated simulation model can also be manually edited by a user to incorporate features not represented in the information models or not supported by the model generation mechanism.

The decision-modeling system also transforms input data in a format suitable for efficient execution of the simulation model. This format is referred to as the modeling techniques' specific data model.

The generated simulation model is executed by a commercially available simulator.

9.3.2 Representation of Supply Chain Entities

The main task is to transform information models into an executable simulation model. To achieve this, the main entities of the supply chain configuration simulation model are defined. There are three main types of supply chain entities in the proposed supply chain simulation model—supply chain units, customers, and products. Links among supply chain units are represented by transportation functions of supply chain units. Several authors have indicated that a supply chain units or a node in the supply chain can be represented in a generic form (Chandra et al. 2000; Pontrandolfo and Okogbaa 1999; Hung et al. 2006). Figure 9.2 shows one representation of the generic supply chain unit. The concept of the generic unit essentially corresponds to supply chain representation in the Supply Chain Operations Reference (SCOR) model. The important aspect of the SCOR model is the presence of both global and local control (i.e., planning processes).

Fig. 9.2 A generic
representation of supply
chain unit

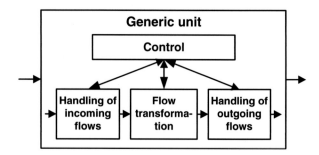

The data model described in Chap. 7 is expanded to include information necessary for describing generic units and supply chain simulation. Therefore, the *SupplyChainUnit* class is shown to be an aggregation of *IncomingFlows, FlowTransformation,* and *OutgoingFlows* classes, which represent generic functionality of supply chain units (Fig. 9.3). The generic functionality is specified by using classes representing particular activities. For instance, products received at a supply chain unit are stored in inventory, described by the *Inventory* class. Classes representing supply chain unit level and supply chain wide control mechanisms are also included. The *Inventory* class describes inventory handling at a particular supply chain unit, while the control classes provide communications among supply chain units and coordinate activities with the unit.

Given that the simulation model is generally used to evaluate fixed supply chain configuration, it is developed on the basis of the object model containing objects, which are instances of classes defined in the class diagram of the generic supply chain unit.

Figure 9.3 also shows classes for representing products and customers. Classes representing global and local control mechanisms are abstract classes and their elaboration is case specific.

9.3.3 Model Generation

The simulation model is generated on the basis of the object diagram. The object diagram contains realizations of classes shown in Fig. 9.3. Realizations are created according to optimization outcomes (for instance, optimization yields that three out of five manufacturing units are to be opened at selected locations; objects representing these three manufacturing units are created) or for a given fixed supply chain configuration to be evaluated.

Different mechanisms are used to represent various entities from the supply chain object model. In the case of using ARENA (Rockwell 2001) as a simulation modeling tool, a supply chain unit object is represented as a standardized sequence of simulation modeling blocks, customer zones are represented using a differently structured sequence of simulation modeling blocks, products and materials are

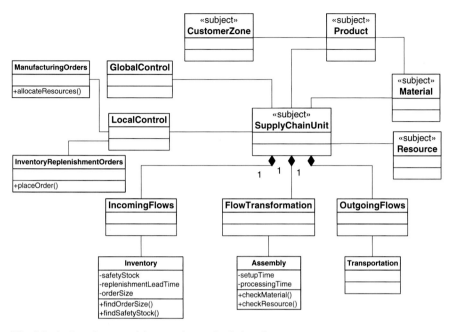

Fig. 9.3 A class diagram of the generic supply chain unit

represented using simulation modeling entities, and resources are represented using the resource module.

An ARENA submodel is generated for every supply chain unit included in the configuration (i.e., for every *SupplyChainUnit* object). Such a submodel for one of the supply chain's units is shown in Fig. 9.4. The *FlowTransformation* object is transformed into a sequence of processes realizing manufacturing order for processing, setting up resources, requesting materials from the stock, and finally assembling the product. The object diagram prescribes that flow transformation is needed and allows the setting of variables in the ARENA model (for instance, the *setupTime* attribute is used to generate a corresponding variable in the ARENA simulation model). At the same time, the object diagram does not specify the flow transformation process. That is perceived as model method and modeling tool specific data, which determine transformation of the object in ARENA blocks. Products and materials are represented by ARENA entities. Arrays are used to deal with multiple products and resources. The model generator in the decision support system is implemented using Visual Basic (VB). It creates ARENA objects using the ActiveX technology (actually, the same data model can be used to create a simulation model in other simulation modeling environments supporting the ActiveX technology).

A separate submodel is used to represent *CustomerZone* objects. This submodel is used to generate customer demand and to serve as a final destination for finished products.

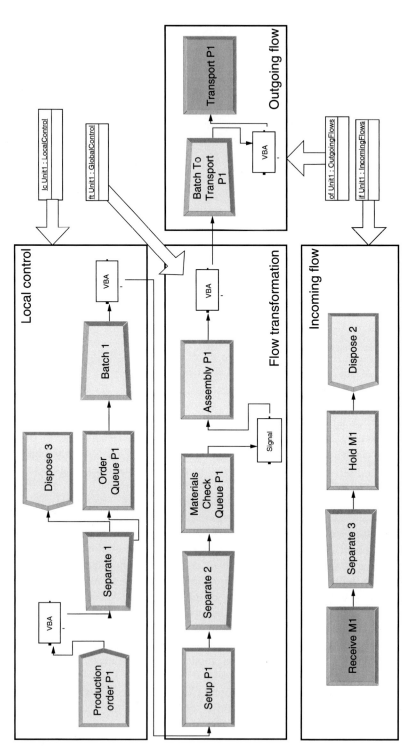

Fig. 9.4 Generation of the ARENA model from the object diagram

The main model generation transformations are summarized as follows:

1. A sequence of simulation modeling blocks is generated for each object of the *CustomerZone* type. This sequence represents generation and queuing of demand orders and receiving.
2. The ARENA resource table is populated by generating an entry for each object of the *Resource* type.
3. An ARENA entity is generated for each object of the *Product* type.
4. An ARENA submodel is generated for each *SupplyChainUnit* object. It consists of four sets of simulation modeling blocks corresponding to objects that compose the *SupplyChainUnit* object. The local control set is generated from the appropriate object of the *LocalControl* type (the object is named *lc_Unit1* in Fig. 9.4). Other sets of blocks are generated in a similar manner.

Global control, and some other local control mechanisms not shown in Fig. 9.4, are included in the supply chain modeling template or manually developed. They are implemented using VB code.

The modeling method specific data model is also generated and populated during the model generation process. The data model organizes data in a manner suitable for execution of the simulation model. This ensures quick access of necessary data items. The data model consists of multiple spreadsheets containing information about structure and operational characteristics of the system. At the beginning of simulation, modeling data from the data model are loaded in the simulation model. Before loading, intermediate data have been created by converting the data model tables from the Excel format into the text format because ARENA reads text files much faster than Microsoft Excel files. Some of the data tables are loaded into ARENA arrays for access by ARENA objects, while some others are loaded in VB arrays for access by control functions.

A more detailed description of functions performed by individual blocks of the simulation model, and structuring of the modeling technique specific data model for a specific industrial case study, can be found in Chandra and Grabis (2003).

The generated simulation model is subsequently used to evaluate the given supply chain configuration. The automated generation enables rapid development of simulation models representing various alternative supply chain configurations.

9.3.4 Sample Simulation Results

As indicated earlier in the chapter, simulation modeling is used to evaluate temporal aspects of supply chain configuration. Sample simulation results illustrate evaluation of the robustness of supply chain configuration in the case of disruptive events (the sample is adopted from the case study of establishing an automotive supply chain in emerging markets (Chandra and Grabis 2002).

Simulation is used to compare two alternative supply chain configurations differing by the number of suppliers. The first configuration (D1) uses MS1 as a

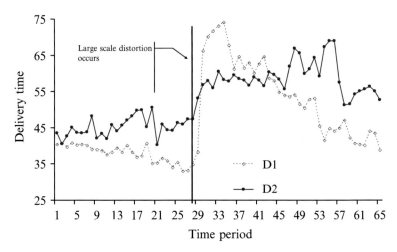

Fig. 9.5 Average delivery time for alternative designs D1 and D2

single steel supplier while the second configuration (D2) uses MS3 and MS5. Among other performance measures, the configurations are compared according to the ability to maintain high delivery time reliability in the case of disruptive events. Figure 9.5 shows delivery time changes in the case of a large-scale distortion, which causes a loss of supplies from some of the second tier suppliers for several periods.

During that time, lost suppliers are replaced by new suppliers. Reported results are obtained over 20 replications, and each replication is 75 periods long. The configuration including only MS1 suffers more intense immediate effect, but further on, it is capable of recovering. The configuration including MS3 and MS5 has difficulties reestablishing previous delivery promptness mainly due to high capacity utilization, which prevents quick replenishing of lost deliveries. Other performance measures, such as inventory costs and total costs, do not change significantly in both cases.

9.4   Hybrid Modeling

Hybrid modeling for supply chain configuration is primarily used for establishing supply chain configuration. Additionally, it can be used in the preselection and results analysis phases of the supply chain configuration methodology. Mathematical programming simulation hybrid models are generally used for the selection phase, while statistical models can be brought into the preselection and evaluation phases. Thus, a single hybrid model can potentially cover multiple configuration phases.

A large number of different hybrid models can be designed. The following subsections describe common principles used in the development of hybrid models and introduce main classes of hybrid models.

9.4.1   General Approach

A hybrid model is part of the integrated supply chain decision-modeling frame-work. It represents a tightly coupled subset of decision-making models. Hybrid modeling starts with semiautomated generation of appropriate optimization and simulation models (Fig. 9.6). These models are generated on the basis of information models, which also provide the means for information exchange between the hybrid model and information sources. The optimization model is used to obtain a

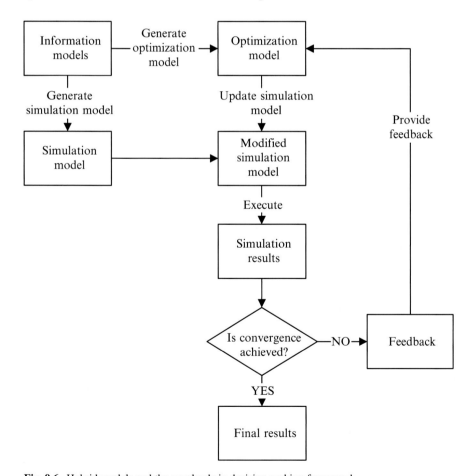

Fig. 9.6 Hybrid models and the supply chain decision-making framework

tentative supply chain configuration. The simulation model is modified according to obtained optimization results. Modifications may range from simple updating of parameters to modification of the supply chain network structure in the simulation model. For instance, if the optimization model suggests opening new warehouses, these should be included in the simulation model. To make such modifications, the automated simulation model development capabilities are essential, otherwise model development and usage becomes too expensive. After simulation modeling is performed, selected criteria are tested for convergence. If convergence has not been achieved, feedback from simulation to optimization is provided and the next optimization iteration is started after feedback information is incorporated into the optimization model. Feedback mechanisms are discussed later in this chapter.

Shanthikumar and Sargent (1983) classify hybrid models by identifying the model that is most important in a pair of simulation and optimization models. There are four classes of hybrid models:

1. Hybrid models whose behavior over time is obtained by alternating between using independent simulation and analytic models.
2. Hybrid models in which a simulation model and an analytical model operate in parallel over time, with interactions through their solution procedure.
3. Hybrid models in which a simulation model is used in a subordinate way for an analytic model of the total system.
4. Hybrid models in which a simulation model is used as an overall model of the total system and requires values from the solution procedure of an analytical model representing a proportion of the system for some or all of its input parameters.

In this chapter, two primary types of hybrid modeling are distinguished in the supply chain modeling context:

1. Sequential hybrid modeling—integration between simulation and optimization is implemented by updating parameters of simulation and optimization models. For instance, a supply chain configuration is optimized using a given processing cost parameter. A simulation model for the configuration established by optimization is executed and the processing cost parameter is updated. The next optimization iteration uses updated values of the parameter. This updated value represents some aspects of supply chain behavior previously not accounted for. One can say that all optimization trials are performed before calling simulation.
2. Simultaneous hybrid modeling—integration between simulation and optimization is implemented by evaluation of objective function. In this case, objective function of the optimization model is not available as a closed form expression (or its analytical evaluation is too complex). An optimization model sets values of decision variables. Simulation modeling results obtained using these decision variables as input parameters are used to find a value of the optimization objective function. The value found is passed back to the optimization model. One can say that simulation is called on each optimization trial. Simulation

based optimization (Carson and Maria 1997) also can be perceived as a type of simultaneous hybrid modeling. Breadth and scope of the simulation model distinguishes these two approaches.

Simulation models often include analytical models for run-time decision-making. For instance, an inventory management algorithm is implemented to make inventory replenishment decisions. Such models closely resemble the fourth type of hybrid model according to the classification by Shanthikumar and Sargent (1983). However, this practice has become so widespread that incorporation of analytical models for performing some specific functions during simulation can be perceived as one of the base simulation modeling capabilities.

9.4.2   Meta-Model Based Feedback

The hybrid models discussed above inherit several drawbacks from their simulation component:

- Computational time substantially depends upon complexity of simulation models.
- Stochastic outcomes complicate feedback to optimization and interpretation of results.

In some special cases, the impact of these limitations can be further reduced by using a hybrid modeling approach proposed by Chandra and Grabis (2005). This approach allows only one-way interactions between simulation and optimization (i.e., simulation results do not depend upon optimization results while optimization uses information obtained using simulation). It uses simulation to estimate the performance of the fixed part of the supply chain subject to dynamic and stochastic factors. Potential supply chain partners are modeled by a set of input parameters characterizing their aggregated properties. A regression based meta-model relating these parameters to their impact on supply chain performance is developed according to simulation results. This meta-model is incorporated as a constraint into a final optimization model, which is used to finalize supply chain configuration. The meta-model is used because simulation yields the performance estimates needed for optimization at only a few discrete points, while the meta-model is a continuous function. Further presentation of this approach is based on a supplier selection case study.

9.4.2.1   Case Description

The supplier selection problem is adopted from Weber et al. (2000). It concerns procurement of one major raw material. Suppliers are selected for the length of a relatively long planning horizon. The number of suppliers to be selected is predefined, and suppliers have restricted supply capacity. There are three supplier selection criteria: (1) material price, (2) delivery reliability measured by percentage of late items, and (3) quality measured by the percentage of rejected items. Supplier data characterizing the selection criteria are given for candidate suppliers. Materials are delivered to the manufacturer in just-in-time mode.

The multi-objective optimization problem is formulated as

$$\min Z = (Z_1, Z_2, Z_3) \tag{9.1}$$

subject to

$$\sum_{j=1}^{N} x_j \geq D \tag{9.2}$$

$$Z_1 = \sum_{j=1}^{N} \rho_j x_j \tag{9.3}$$

$$\sum_{j=1}^{N} v_j = P \tag{9.4}$$

$$x_j \leq v_j w_j^{\max} \tag{9.5}$$

$$x_j \geq v_j w_j^{\min} \tag{9.6}$$

$$Z_2 = \sum_{j=1}^{N} \lambda_j x_j \tag{9.7}$$

$$Z_3 = \sum_{j=1}^{N} \beta_j x_j \tag{9.8}$$

$$v_j = \begin{cases} 0, \text{if } x_j = 0 \\ 1, \text{if } x_j > 0 \end{cases}, \quad j = 1, \dots, N \tag{9.9}$$

Z_1 represents the material price criterion, Z_2 represents the delivery reliability criterion, and Z_3 represents the quality criterion. x_j is the quantity to be purchased from the jth supplier. N denotes the number of candidate suppliers. v_j indicates whether the jth supplier is included in the supply network. $w_j^{\min}$ and $w_j^{\max}$ limits minimum and maximum total order quantities for the jth supplier, respectively. ρ_j is the net purchase price per unit for jth supplier, λ_j is the percentage of late items for the jth supplier, β_j is the percentage of items rejected for the jth supplier.

9.4.2.2 General Approach

The standalone optimization of this supplier selection problem is complicated by different dimensions of supplier selection criteria. The simulation model is used for evaluating the impact of delivery performance and quality of materials on manufacturing costs. It represents the actual manufacturing system at an appropriate level of abstraction. If all selection criteria are expressed in terms of the manufacturing costs incurred, then the multi-criteria problem is transformed into a single objective optimization problem, which selects suppliers in such a way that minimizes the costs associated with sourcing from these particular suppliers. To perform simulation without knowing specific suppliers, these specific suppliers are abstracted by an aggregated supplier.

Alternatively, the optimization of a supply network could be performed using the simulation model directly or using a hybrid optimization–simulation model, which invokes the simulation model on each optimization trial, and is computationally challenging.

The hybrid modeling process, using the meta-model as a feedback mechanism, involves the following steps:

1. Identify supplier selection criteria and a plausible range of values for these criteria.
2. Develop the simulation model of the manufacturing system and experimental design.
3. Run simulations.
4. Construct the meta-model relating suppliers' characteristics and manufacturing costs.
5. Integrate the meta-model into the optimization model.
6. Optimize the supply network.

During Step 1 of the modeling process, a list of supplier selection criteria is compiled. Plausible ranges of criteria are also identified. For instance, the percentage of defective materials varies from 1 to 5 %. A simulation model of the manufacturing system is developed at Step 2. Manufacturing operations are simulated without specifying suppliers, and an abstract aggregated supplier is used instead. The importance of experimental design is stressed because simulation experiments need to cover the entire range of plausible values in the supplier selection criteria. During the simulation, manufacturing costs are evaluated subject to the suppliers' characteristics corresponding to the chosen criteria. Simulation modeling results are used to construct a regression-based meta-model. This meta-model takes the suppliers' characteristics as input parameters and returns the total manufacturing cost. The meta-model can be readily integrated into the supplier selection optimization model. The optimization models select suppliers and determine order quantities from each supplier by minimizing the calculated manufacturing costs using the meta-model and purchasing costs, such as material prices included only in the optimization model.

9.4.2.3 Specific Models

The supplier selection problem description defines the list of supplier selection criteria. The minimum and maximum values of supplier data that characterize the selection criteria constitute lower and upper bounds on the range of plausible values, respectively.

Simulation model: Development of the manufacturing system's simulation model is the next step of the modeling process. Simulation modeling of the manufacturing system generally requires more information than needed, just for mathematical programming purposes. Therefore, the following assumptions about the system are made:

1. Procured materials are used to manufacture a single end product.
2. The planning horizon is 1 year, divided into 8-h normal work days.
3. Demand for the end product is uniformly distributed across the planning horizon (this assumption is consistent with the just-in-time system condition).
4. The manufacturing system has the fixed and relatively high capacity utilization level.
5. Materials are supplied at the beginning of each day in a single batch.
6. Each day, a percentage of all items is delivered late. This percentage is distributed according to the logarithmical normal distribution with mean λ^* and standard deviation $g\lambda^*$. Processing of items delivered late can be started only after all on-time deliveries are processed.
7. Each day a percentage of all items are rejected because of poor quality. This percentage is distributed according to the logarithmical normal distribution with mean β^* and standard deviation $h\beta^*$. The rejected items can be processed on the next day, after all on-time deliveries are processed. Processing of these items takes ~2.5 times longer than processing of standard items. This increase in processing time depends upon the quantity of late deliveries on a particular day.
8. Items delivered on time can be processed within normal working hours. However, processing of late deliveries and rework can extend to overtime. The premium charge B \$/item is imposed for overtime processing (rework counts with respect to increased processing time).
9. Overtime cannot exceed 2 h. If more work is left, it is carried over to the next day.

For selection purposes, only suppliers' characteristics dependent manufacturing costs *MC* need to be accounted for. In this case, these are only overtime costs, which are computed as overtime cost B times the number of items processed in overtime.

The experimental design is created by varying values of the aggregated suppliers' characteristics λ^* and β^* within specified ranges. Multiple replications of simulation modeling are executed for each experimental cell.

Meta-model: The manufacturing cost *MC* is the output variable of simulation modeling. λ^* and β^* are input variables of simulation modeling. The results of simulation modeling are used to develop a regression-based meta-model relating manufacturing costs to suppliers' characteristics. In this case, a linear meta-model is used

$$MC = a_0 + a_1\lambda^* + a_2\beta^* \tag{9.10}$$

where $a_k, k = 0, .., 2$ are estimated coefficients of the meta model (see Kleijnen (2005) for more details on creating meta-models from simulation results).

Optimization model: After developing the meta-model, the supply network is established using mathematical programming. The meta-model yields suppliers' characteristics dependent manufacturing costs. The mathematical programming model also includes other strategic purchasing costs. In this case, these are net purchasing costs per item. Because the simulation model expresses the delivery performance and quality criteria in terms of manufacturing costs, the single objective mathematical program corresponding to the multi-criteria supplier selection problem described earlier is formulated.

The objective function of this mathematical program is

$$\min TC_m = Z_1 + C \qquad (9.11)$$

where TC_m are total costs, accounting for costs due to the impact of delivery performance and quality.

It is evaluated, subject to constraints in Eqs. (9.3–9.7) and Eq. (9.10). The additional constraint representing the manufacturing costs is introduced per Eq. (9.12).

$$C = \sum_{j=1}^{N} \left[\left(a_0 + a_1 \lambda_j + a_2 \beta_j \right) x_j D^{-1} \right] \qquad (9.12)$$

This constraint is derived from the meta-model. The simulation model, and consequently the meta-model, uses a single aggregated supplier. To account for the individual contributions of each supplier, the manufacturing cost is weighted by a proportion of items purchased from a particular supplier.

9.4.2.4 Results

Experimental studies are conducted to demonstrate the proposed supplier selection approach. These compare the supplier selection results obtained using the proposed multi-objective approach, and the single-objective optimization with respect to purchasing price only.

To assess the impact of assumptions, different sets of experiments are conducted as per the criteria defined in Table 9.2. The overtime processing cost related to the material purchasing price is chosen. g characterizes the level of uncertainty about how many items are delivered late each day. h characterizes the level of uncertainty about how many items are rejected each day. The total demand is 40,000 items per day. The capacity utilization rate is 95 %.

Table 9.2 Sets of parameters of the manufacturing system considered

	G	h	B
1st set	1	1	0.2
2nd set	0.2	0.2	0.2
3rd set	1	1	2
4th set	0.2	0.2	2

Table 9.3 Values of aggregated suppliers' characteristics used

	λ^*	β^*
Upper level	1	0.2
Lower level	7	2.3

Table 9.4 Supplier selection results in '00000

	Multi-objective								Single objective
	TC_m	x_1	x_2	x_3	x_4	x_5	x_6	TC_m^*	TC_1^*
1st set	29.0	36.9	107.9	0.4	0.4	0	0.4	28.8	28.9
	($\pm$0.02)							($\pm$0.02)	($\pm$0.02)
2ed set	28.8	36.9	107.9	0.4	0.4	0	0.4	28.7	28.7
	($\pm$0.02)							($\pm$0.0)	($\pm$0.00)
3rd set	31.1	0.4	30.3	107.9	7.0	0	0.4	31.3	38.5
	($\pm$0.02)							($\pm$0.02)	($\pm$0.13)
4th set	30.7	0.4	46.7	98.1	0.4	0	0.4	30.7	36.6
	($\pm$0.01)							($\pm$0.00)	($\pm$0.05)

Note: 95 % confidence intervals are provided in parenthesis

The supplier selection problem is solved for each set of parameters of the manufacturing system. The full factorial experimental design needed to develop the meta-model is created for each set. It is developed by pairing the upper and lower values of λ^* and β^* identified in Table 9.3. Fifty replications are executed for each experimental cell, and a meta-model for each set is fitted. The R-squared values for all meta-models exceed 0.85.

The fitted meta-models for each set of parameters are incorporated separately into the optimization model to make the supplier selection decisions. The optimization model uses values of net purchase price per unit ρ_j, percentage of items late λ_j, and percentage of items rejected β_j, as given in Weber et al. (2000, Table 1). The number of candidate suppliers is six. The number of suppliers to be selected is five. The quantity of raw materials allocated to each supplier is reported in Table 9.4. The table also compares the total cost TC_m as estimated using optimization model, the total cost TC_m^* obtained by simulating the optimized supply network, and the total cost TC_1^* obtained by simulating the optimized supply network, if $a_k = 0$, $k = 0, .., 2$ (i.e., only the price criterion is used). In the case of $a_k = 0, k = 0, .., 2$, the optimization model yields $x_j = \{36.9, 107.9, 0, 0.4, 0.4, 0.4\}$. The results show that differences between supplier selection results for the 1st and 2nd sets are not statistically significant. Additionally, the multi-objective problem-solving approach and the single-objective approach yield statistically identical results. The overtime driven manufacturing cost is too low to substantially affect supplier selection decisions. The majority of items are ordered from the supplier offering the lowest unit price, despite its poor on-time delivery performance. However, if the overtime cost factor B is set at a high level (3rd and 4th sets), the manufacturing cost substantially influences supplier selection decisions. The majority of items are ordered from the supplier offering the highest price and proving the good on-time delivery performance and quality. TC_1^* is substantially worse than TC_m^* for the 3rd and 4th sets, highlighting the importance of accounting for not only the purchasing price factor but for other factors as well.

The above-described approach can be expanded to other supply chain configuration problems in addition to the supplier selection problem. However, it is more suitable, if the supply chain structure is partially fixed.

9.5 Summary

Simulation modeling is a well-established tool for evaluation of fixed supply chain configurations. This chapter focuses on making simulation an integral part of the supply chain configuration methodology. This is achieved by proposing the automated model development approach, which advocates generation of simulation models on the basis of information models and optimization outcomes. Automated model development is beneficial because the simulation model does not need to be manually redeveloped for every new configuration.

There are several limitations that need to be addressed. The proposed approach allows for automated development of structure of simulation models while information models driven development of control mechanisms is supported to a limited extent. Improving implementation of control mechanisms is a direction of further research. The approach can only be efficient if high reusability is achieved. Therefore, a precise specification of model transformation and generation mechanisms is required.

Hybrid modeling is a powerful technique proving additional opportunities for comprehensive supply chain configuration modeling. It implements the principle of the model synergies. However, there are several major issues to be addressed. These are the development of a theoretically sound feedback mechanism, accumulation of general knowledge on application of hybrid modeling, and expansion beyond joining simulation and optimization modeling.

Ad hoc feedback mechanisms for the sequential approach are available for problems like production planning with continuous variables. However, there are profound difficulties in implementing the feedback mechanism for the supply chain configuration problems involving binary variables. Additionally, convergence properties of algorithms used for solving hybrid models are sparsely investigated. That leads to uncertainty concerning the quality of solutions.

Hybrid models tend to be developed on a case-by-case basis. Therefore, little guidance for developing hybrid models is available. The problem is also important for simulation-based optimization models. Despite the availability of advanced computational tools, empirical evidence of application of these tools in combination with complex discrete event simulation models is limited.

Hybrid modeling was originally applied to the integration of simulation and optimization models. However, integration with other quantitative models is also possible. This direction is particularly relevant for the supply chain configuration problem because statistical models are frequently used at the preselection stages. Currently available models consider a linear chain of statistical models used for preselection, and optimization models used for selection (and simulation models used for evaluation). Alternatively, a sequential hybrid model could be constructed for iterative evaluation. For instance, a statistical preselection model captures suppliers' commitment strengths, which is influenced by a purchasing quantity determined using optimization.

Finally, this chapter emphasizes that the common information basis, and the automated development and modification of hybrid models are essential for efficient application of these models.

References

Bowersox DJ (1972) Planning physical distribution operations with dynamic simulation. J Market 36:17–25

Byrne MD, Bakir MA (1999) Production planning using a hybrid simulation – analytical approach. Int J Prod Econ 59:305–311

Carson Y, Maria A (1997) Simulation optimization: methods and applications. In: Andradóttir S et al (eds) Proceedings of the 1997 of the winter simulation conference, Atlanta, pp 118–126

Chandra C, Grabis J (2002) Modeling floating supply chains. Proceedings of eleventh annual industrial engineering research conference, Orlando, USA, Accessed 19–22 May 2002

Chandra C, Grabis J (2003) A data driven approach to automated simulation model building. Proceedings of the 18th European simulation symposium, ESS2003, Delft, The Netherlands, pp 372–380, Accessed 26–29 October

Chandra C, Grabis J (2005) Multi-objective supplier selection using simulated cost estimates, Proceedings of European modeling and simulation symposium, EMSS' 2005, pp 143–150

Chandra C, Nastasi AJ, Norris TL, Tag P (2000) Enterprise modeling for capacity management in supply chain simulation. Proceedings of ninth industrial engineering research conference, Cleveland, Accessed 21–24 May

Hung WY, Samsatli NJ, Shah N (2006) Object-oriented dynamic supply-chain modelling incorporated with production scheduling. Eur J Oper Res 169:1064–1076

Ingalls RG (1998) The value of simulation in modeling supply chains. In: Medeiros DJ, Watson EF, Carson JS, Manivannan MS (eds) Proceedings of the 1998 winter simulation conference. Washington DC, pp 1371–1375

Kleijnen JPC (2005) An overview of the design and analysis of simulation experiments for sensitivity analysis. Eur J Oper Res 164:287–300

Law AM, Kelton WD (2014) Simulation modeling and analysis. McGraw-Hill, New York

Mertins K, Rabe M, Jakel FW (2005) Distributed modelling and simulation of supply chains. Int J Comput Integrated Manuf 18:342–344

Nolan RL, Sovereign MG (1972) A recursive optimization and simulation approach to analysis with an application to transportation systems. Manag Sci 18:676–690

Pontrandolfo P, Okogbaa OG (1999) Global manufacturing: a review and a framework for planning in a global corporation. Int J Prod Res 37:1–19

Rockwell Software (2001) ARENA: user's guide. Rockwell Software Inc., Sewickley

Shanthikumar JG, Sargent RG (1983) A unifying view of hybrid simulation/analytic models and modeling. Oper Res 31:1030–1052

Simchi-Levi D, Kaminsky P, Simchi-Levi E (2003) Designing and managing the supply chain. McGraw-Hill/Irwin, New York

Vamanan M, Wang Q, Battab R, Szczerba RJ (2004) Integration of COTS software products ARENA & CPLEX for an inventory/logistics problem. Comput Oper Res 31:533–547

Van der Vorst JGAJ, Beulens AJM, Van Beek P (2000) Modelling and simulating multi-echelon food systems. Eur J Oper Res 122:354–366

Weber CA, Current J, Desai A (2000) An optimization approach to determining the number of vendors to employ. Supply Chain Manag 5:90–99

Part III
Technologies

Chapter 10
Information Technology Support for Configuration Problem Solving

10.1 Introduction

Supply chain management and information technology are tightly coupled. Implementation of supply chain strategies would be very difficult without the support of information technology. At the same time, many developments in information technology have arisen from requirements set by enterprises seeking collaboration with their partners in the supply chain environment. For instance, the development of web services has been driven by a need for flexibility and more open information systems interfaces to support reconfigurability.

This chapter describes the information technologies used to support supply chain configuration, both from the decision-making and decision-implementation perspectives. Information technologies include hardware, communications, and software. Here, the focus is on software. Hardware and communications are described only at the conceptual level in relation to the general architectures of information technology solutions in the supply chain management framework.

The chapter starts with outlining major aspects of information technology usage in supply chain management, and supply chain configuration in particular. The classification of applications used is provided. Section 10.3 describes applications used for decision-making purposes. Section 10.4 discusses the supply chain management information system, with emphasis on the use of flexible, service-oriented architecture. A prototype of the supply chain configuration decision-making system is described in Sect. 10.5.

© Springer Science+Business Media New York 2016
C. Chandra, J. Grabis, *Supply Chain Configuration*,
DOI 10.1007/978-1-4939-3557-4_10

10.2   Requirements

Information technology is irreplaceable, primarily because of the amount and complexity of information processing and the physical distribution engaged by each supply chain member.

To describe information technology support for supply chain management, and configuration in particular, it is important to distinguish between information technology support for decision-making needs and information technology support for supply chain execution according to the decisions made. In the former case, applications of information technology include providing input data to decision-making models, solving decision-making models, and presentation of results. In the latter case, applications of information technology include communicating data among supply chain units, processing of transactions, online decision-making and monitoring of supply chain efficiency.

Several key requirements for the information technology solutions used to support supply chain management include the following:

- Horizontal and vertical integration. Supply chain management processes are executed across functional domains within individual units, as well as across enterprises involved in the supply chain network. Sharing of data and processes should be supported at a level desirable for a particular decision-making problem. Its implementation should enable relatively simple replacement of components comprising the architecture of the supply chain information system.
- Security. Effective supply chain management depends upon a level of trust between partners, especially in the case of rapidly evolving supply chain structures. Clearly defined and strictly enforced security policy has an important role in trust building. Obviously, information technology solutions should also support all common data security requirements.
- Reliability. A high level of information systems availability is required to support collaborative decision-making and implementation of decisions.
- Scalability. The intensity of product and information flows can change quickly, along with changes made in the supply chain configuration. Information technology solutions should be able to accommodate these changes.

Functional requirements for supply chain information systems depend upon a particular supply chain configuration problem (some of the common functional requirements are identified throughout the book). These requirements, along with earlier defined key requirements, determine the design of the supply chain information system. Several common features of such a design can be identified. Information technology solutions are made up by pairing hardware, communications, and software. Typical characteristics of such solutions in the supply chain environment include:

- Heterogeneity. Information technology solutions are heterogeneous, both within an enterprise and throughout the supply chain. The internal heterogeneity is caused by using different applications for solving various supply chain management problems. For instance, decision-making is performed using an advanced planning system while decisions are implemented using the Enterprise Resource Planning (ERP) system of another supplier because an enterprise attempts to use best-of-breed solutions. The external supply chain heterogeneity is caused by supply chain members using different information technology architectures, which are still strongly influenced by local tradition. It is important to note that heterogeneity is characteristic to all levels of information technology solutions. At the logical level, different data models and process representations are used (see Chap. 7). Different software packages are used at the implementation level and different platforms and a means of communication are used at the infrastructural level.

- Distributed system. This characteristic mainly owes to the spatially distributed nature of supply chains. Supply chain partners are physically distributed around the world, and the design of the supply chain information system accounts for this problem. An additional important feature is the lack of centralization that complicates management of the supply chain information system. For instance, one supply chain partner generally is not able to use its system to enforce the security policy in the system owned by another partner.

- Use of public communication channels. Reconfigurable supply chain management information systems use public communication channels and the Internet, in particular, extensively. These communication channels offer a substantially higher degree of flexibility and lower costs while modern technologies can provide an adequate level of security.

- Pervasive computing. Pervasive computing implies that computation can take place on different platforms and in different location. Besides information processing in traditional client–server environment, computations can take place on mobile devices and a wide range of sensors can be incorporated. The data processing operations can be performed through a variety of interfaces such as web, machine-to-machine, and others. In the case of supply chain configuration problems, this feature facilitates data availability and supports collaborative decision-making. The configuration decisions can be made by having either physical or virtual presence at different supply chain locations.

10.3   Types of Technologies

Different types of information systems are used in supply chain management. Information systems involved in supply chain configuration decision-making support and implementation can be classified as follows:

- General-purpose decision modeling applications. These applications are used to develop and solve different types of decision-making problems, including the supply chain configuration problem. Typical representatives include the optimization package LINGO and the simulation package ARENA.

- Problem-oriented decision modeling applications. These packages include specific solutions for particular decision-making problems—in this case supply chain configuration. Examples of such packages are LOCOM[1] and modules of JDA.[2]
- Integrated modeling environments. Decision-modeling applications are supplemented with different service modules, primarily for data handling, management of experiments, and presentation of results.
- Advanced planning systems. Integrated enterprise-level planning systems supporting hierarchical decision-making.
- Data management systems. Special purpose data storage and presentation systems integrating data from various sources to support decision-making. Data warehouses and business intelligence solutions form the backbone of data management systems for decision-making while big data processing oriented solutions are introduced for capturing newly arising data analytics opportunities in a flexible manner.
- Enterprise applications. Legacy systems, ERP systems, specialized applications and service-oriented applications ensure transactional data processing and business process execution. They serve as data sources and consumers to decision-making applications; mainly custom-built systems supporting specific functional areas of enterprises. Legacy systems are developed on the basis of outdated computing platforms.
- Workflow management systems and groupware. These systems are used to support collaborative decision-making and implement decisions made.
- Web services and web applications. These technologies support supply chain wide sharing of information and processes. They provide a flexible solution for exposing data and functionality to supply chain partners.
- Information modeling packages and integrated development environments. These systems are used for information systems development purposes, supporting both information systems modeling and actual implementation.

The list of information systems does not include software packages used at the infrastructural level, such as server management systems and network routing systems.

The specialized applications differ from the ERP systems by focusing on particular kind of enterprise and supply chain processes. Sample specialist applications are transportation management systems, warehouse management systems, and supply chain management systems (dealing with inventory and materials flow management) (Helo and Szekely 2005). Nyman (2012) classifies supply chain applications as those supporting data management, data exchange, data tracking, and process execution.

---

[1] http://www.logistics-designer.com/.
[2] http://www.jda.com/.

Despite importance of coordination and collaborative planning in supply chains, a majority of companies still restrict their planning activities to the intra-company level (Fuchs and Otto 2015). That is often explained by lack of adequate information and communication technologies to support collaboration and specifically for collaborative decision-making in decentralized supply chains (Hernández et al. 2014). Qu et al. (2010) establish a dedicated service platform for decentralized decision-making in supply chains. The platform includes atcPortal, providing a common point-of-control for supply chain configuration decision-making. It uses a standardized set of messages and interfaces to exchange supply chain configuration data. Agostinho et al. (2015) conceptualize integration of information systems in dynamic networks. They emphasize importance of knowledge management and model-driven development technologies in achieving sustainable interoperability.

10.4 Supply Chain Integration

An information technology solution for supporting supply chain configuration is built according to the supply chain framework and the architecture of the decision support systems proposed in Chaps. 4 and 5, respectively. It ensures integration among

- Supply chain units.
- Different applications involved in the supply chain configuration process.
- Different types of information flows.

10.4.1 Overviews

The supply chain consists of interrelated units each possessing their own supply chain management information systems and decision support systems, where the former are mainly used for supply chain execution, and the latter for planning and decision-making at various levels of supply chain management (Fig. 10.1). One of the units or their group is considered as a focal supply chain unit, which drives the supply chain configuration initiative.

The units have appropriate interfaces for integrating various types, if information flows across the supply chain. They exchange data used for decision-making (DM), reporting (R), and execution (E) purposes. The decision-making data exchange interface is used to exchange data required for decision-making as well as receiving decision-making results. For instance, a supplier provides the expected material price information as a decision-making input and receives back the expected order quantity. This information is received at and provided by a supply chain configuration hub, which coordinates decision-making at the supply chain level. The execution data exchange interface ensures information exchange among

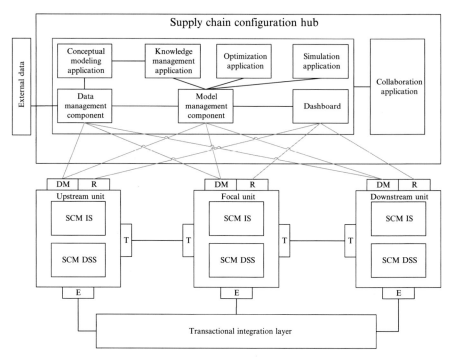

Fig. 10.1 Technological support for supply chain integration: supply chain configuration perspective

the supply chain units after the configuration decisions have been implemented. For example, actual purchasing orders are exchanged among the supply chain units. The configuration decisions affect the execution because data connections should be established and requirements and restrictions imposed by the configuration decisions should be taken into account. The execution data are exchanged at the transactional (T) integration layer which supports various data exchange methods in both centralized and decentralized modes. The reporting data exchange interface ensures that data characterizing performance of the supply chain configuration decisions are gathered and sent to the supply chain configuration hub. The hub receives data already in an aggregated form prepared by the individual supply chain units, specifically for supply chain configuration evaluation purposes. For instance, rather than sending all purchasing information, the supplier sends only the average actual materials sales price.

The supply chain configuration hub is a temporal or permanent information system developed for driving and coordinating the supply chain configuration initiative. It is operated by the focal supply chain unit or a group of supply chain units. Depending on the level of information sharing and collaboration among the units, the supply chain configuration hub can be operated jointly by all supply chain units or separately by parties concerned. That directly affects the decision-making scope and impact on supply chain execution.

The supply chain configuration hub performs data integration activities for supply chain configuration decision-making purposes and manages decision-modeling and decision-making processes. It contains all components required by the supply chain configuration methodology. Different applications used at the supply chain configuration hub are not necessarily deployed at a centralized location but can be invoked from the units' decision support systems. The hub also can be perceived as the supply chain configuration decision support system at the supply chain level while individual supply chain units operate their own decision support systems within their restricted scope.

10.4.2 Decision-Modeling Components

The supply chain configuration hub provides means for collaborative exploration of the configuration problem from various views on the basis of the common information and knowledge basis. Ready-to-use software packages are used for implementation of many of the components. The key functions of the components are listed in Table 10.1.

Table 10.1 Key functions of the components of the supply chain configuration hub

Component	Functions
Model management component	Experimental design Execution of experiments Orchestration of modeling activities Maintenance of modeling results
Conceptual modeling application	Creation and maintenance of conceptual models
Data management component	Offline data retrieval On-demand data retrieval Data maintenance Data transformation Data sharing
Knowledge management application	Maintenance of common dictionary Maintenance of best practices Maintenance of modeling templates
Optimization application	Optimization model formulation Model solving Analysis of modeling results
Simulation application	Simulation model development Simulation Analysis of simulation results
Dashboard	Reporting of modeling objectives and measures Visualization of modeling results Visualization of monitoring data
Collaboration application	Communication among parties involved Sharing of modeling and decision-making artifacts

The supply chain strategy updating step of the supply chain configuration methodology is supported primarily by the knowledge management application and the dashboard. The knowledge management system is used to create informal supply chain strategy and configuration scope documents while configuration targets are setup in the dashboard. The conceptual modeling application supports data and process modeling techniques. The conceptual models developed in the second step of the supply chain configuration methodology are created using the generic models also maintained by the conceptual modeling application. The models created are exported in a standardized model exchange format such as XMI (Lundell et al. 2006). The data management component is responsible for providing actual data to all decision-modeling components. A data structure is created inside the data management component according to the results of conceptual modeling and it is populated with data gathered from multiple sources. Data are retrieved using traditional ETL approaches (Dolk 2000), or on-demand data interaction as described in Chap. 11. Data transformations are also performed within this component. These are specified within the ETL processes, using XSLT or hardcoded. The data management component maintains the common data structure with data and data should be transformed in multiple formats used by decision-modeling applications.

The model management component is a central component responsible for integration among the components of the supply chain configuration hub and managing the supply chain configuration decision-modeling activities. Supply chain configuration modeling is performed for different scenarios and using different decision-modeling models. The model management component creates and maintains a list of experiments to be conducted. Each experiment is conducted using a specific set of input parameters and a specific modeling method (e.g., optimization or simulation). It executes the experiments and accumulates experimental results. In order to execute the experiments, it receives data from the data management component in a format required by the decision-modeling applications. The experimental results are stored by the data management component and the selected results are shared among the supply chain units using the data management component.

Decision-modeling is performed with optimization and simulation applications as well as other applications which might be used for specific purposes. These are typically packaged commercially available software packages, such as LINGO and ARENA for optimization and simulation, respectively. They are used for model building as well as for model execution. Applications are invoked by the model management component, which also channels the data necessary for modeling purposes.

Integrated decision modeling environments combine multiple functions of the supply chain configuration hub. These can be used to implement some parts of model management, decision-modeling, and data management. Integrated decision-making environments provide model development productivity tools, data management functions and functions for executing experiments and presenting decision-making results. Integrated decision-making environments often are developed as a set of utilities supplementing existing decision-modeling applications. For instance, ILOG provides OPL Development Studio for developing ILOG CPLEX optimization models.

The dashboard and collaboration application are primary applications used for decision-making and monitoring activities of the supply chain configuration. The dashboard provides means for communicating decision-making results and information for evaluation of these decisions during supply chain execution. It lists decision-making objectives and expected values of performance measures as calculated by the supply chain configuration models and compares these with actually observed values reported by the supply chain units. The dashboard also provides supply chain visualizations, including geographical information systems based visualizations. The collaboration application facilitates communication among parties involved in the supply chain configuration process.

Advanced planning systems also provide many functions of the supply chain configuration hub. These are used to plan supply chain execution at all decision-making levels, spanning from the strategic level to the operational level (Stadtler and Kliger 2005). Strategic network planning, or supply chain configuration is the starting point of the advanced planning process (Fleischmann and Koberstein 2015). It provides input to all other planning processes. Advanced planning systems such as SAP APO are often tightly integrated with ERP systems. They require centralization of supply chain decision-making data and are less open for collaborative decision-making activities and usage of specialized tools for some of the supply chain configuration activities. A highly specialized form of advanced planning systems is supply chain design suites, such as Llamasoft.[3] These tools have functions for heterogeneous data integration, cloud based collaboration, and advanced supply chain visualization.

10.4.3 Integration with Execution Components

Execution components or supply chain information systems are used for data processing during day-to-day supply chain operations. They have multiple responsibilities in relation to supply chain configuration:

• Providing information for decision-making.
• Providing information for monitoring.
• Executing supply chain operations according to the decision-making results.
• Integration of online decision-making components into the overall supply chain information system.
• Integration of new supply chain units in the supply chain information system.

---

[3] http://www.llamasoft.com/.

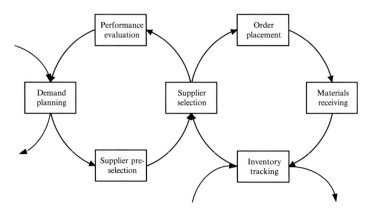

Fig. 10.2 The closed loop interactions between transaction processing and decision-making

The supply chain management information system and decision-making components are tightly interrelated. The decision-making components use information from various sources in the supply chain management information system. While decision-making results are often represented in the supply chain information system in an offline manner; on a case-by-case basis, some of the decision-making results such as operational level supplier selection decisions are made in real-time by invoking decision-making components. This way, there are closed-loop interactions between supply chain execution and supply chain configuration decision-making, as illustrated in Fig. 10.2.

The right-side of the loop in Fig. 10.2 describes the inventory replenishment processes. In this case, suppliers are dynamically selected on a case-by-case basis. Suppliers are selected according to decision-making rules elaborated during the supply chain configuration studies and implemented in an integrated manner. The left-side loop describes the supplier selection processes within the decision-making component of the supply chain information system. Demand planning and inventory tracking are also involved in some other transaction processing or decision-making processes. The supply chain information system is required to support this kind of interactions.

Data exchange with the supply chain configuration hub occurs through the decision-making and reporting interfaces. The transactional data processed by the supply chain information systems are pre-aggregated by individual supply chain units before sending them to the hub. The decision-making results are incorporated in the supply chain information systems either by updating their configuration, setting-up appropriate master data values or invoking decision-making components in real time. Setting a lot-size parameter in ERP systems is an example of the master data setup. The decision-making components invoked in real time can be either implemented as part of a transactional system or provided by the decision support system. The latter mode ensures separation of concerns and allows for easier modification (Zarghami et al. 2012). Web services and component technologies are used for integration of the decision-making components (Siau and Tian 2004).

The transactional integration layer manages information exchange among the supply chain units. A variety of technologies are used to implement this layer. In the case of decentralized supply chain, there is no common infrastructure though focal units often maintain service bus or data exchange hubs (Li et al. 2010). The cloud based integration is an emerging approach to supply chain integration (Radke and Tseng 2015). Cloud computing can be used to support various interoperability scenarios, for instance, interactions among customers through a community cloud or using a private cloud for internal resource planning (Mezgár and Rauschecker 2014).

Flexibility of the integration layer directly affects the ability to incorporate new supply chain units in the supply chain. Standardization and usage of the open web based technologies are primary enablers of rapid integration in the face of supply chain configuration. For example, web services and XML based standards such as ebXML[4] and RosettaNet[5] are widely used (Nurmilaakso 2008; Ahn et al. 2012).

Timely data exchange is enabled by tracking technologies shown in Fig. 10.1 as tracking interfaces. The tracking interfaces are responsible for gathering real-time information about movement of products along the supply chain. Tracking technologies include Near-field communication devices (NFC), Internet, Radio Frequency Data Capture (RFDC), Point-Of-Sales (POS) scanners and bar code scanners (Bowersox et al. 2012; Angles 2005). Communication technologies, such as Global Positioning System (GPS), Global System for Mobile Communications (GSM) and Wireless internet enable to capture and transmit information about products' transportation (Kwak et al. 2014; Butner 2010). The tracking data are used by supply chain management information systems and provide raw data for supply chain configuration reporting.

10.5 Prototype of a Decision-Modeling System

The prototype decision-modeling system implements some of the ideas explained above, as well as those represented in the architecture of the decision-modeling system (Chap. 5).

Figure 10.3 shows components of the prototype of the decision-modeling system. The core part of the decision-modeling system is implemented on the basis of Microsoft Excel, using Visual Basic for Applications. It provides functionality to execute decision modeling processes and maintains data needed for decision-modeling applications. Microsoft SQL Integration Services are used to channel data from data sources to the core part of the decision-modeling system, where these data are arranged in a format suitable for efficient use in decision-modeling applications. Enterprise Architect is used to develop a supply chain process model.

---

[4] www.ebxml.org.

[5] www.rosettanet.org.

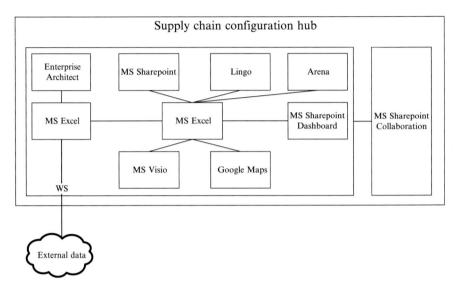

Fig. 10.3 Components of the prototype decision-modeling system

Data about candidate units are extracted from this process model. The decision-modeling applications are LINGO and ARENA for supply chain configuration optimization and evaluation of the established configuration, respectively. The core part of the decision-modeling system invokes LINGO by using its application programming interface, and ARENA is invoked using its COM interface. Automated generation of optimization and simulation models is supported.

Figure 10.4 shows a fragment of a user interface of the decision-modeling system. The right side of the figure contains the optimization model automatically generated from the input data on the basis of a predefined template. The main functions of the prototype decision support system are as follows:

- Development of descriptive supply chain configuration process models (Sect. 5.5.3).
- Extraction of data needed for decision-modeling purposes from the descriptive models, and maintenance of modeling method specific data (Sect. 5.5.3).
- Generation and execution of supply chain configuration optimization models (Sect. 5.5.5).
- Generation and execution of supply chain configuration simulation models according to the optimization results (Sect. 5.5.3).
- Accumulation of decision-modeling results in a format suitable for conducting further analysis (Sect. 5.5.6).

This prototype decision-modeling system has been applied in several supply chain management studies reported in Part IV of this book. The main advantages provided by the system include the following:

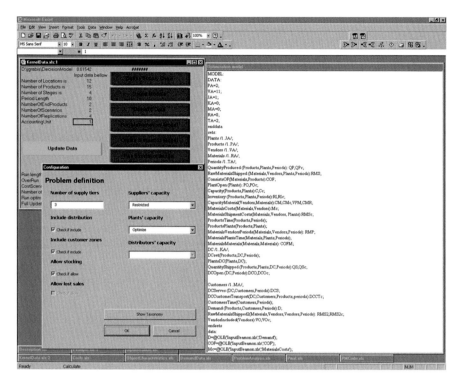

Fig. 10.4 User interface of the prototype decision-modeling system

- The unified data source for supply chain configuration optimization, and simulation models in the form of the supply chain process model, which provides a business, user-friendly description of the supply chain.
- Reduced model development efforts and improved modeling consistency through automated model generation.
- Tight integration between supply chain optimization and simulation models enabling comprehensive appraisal of supply chain configuration decisions.
- Integrated environment for conducting supply chain configuration studies.

10.6 Summary

The supply chain configuration decision-modeling system and the supply chain management information systems consist of a large number of interrelated components. Modern integration technologies are used to bind these components together in a flexible manner. Obviously, neither the decision-modeling system nor the supply chain management information systems are designed to specifically deal with the supply chain configuration problem. The objective is to design such an information technology solution that decision-modeling and implementation systems can be easily designed on the basis of available information technology infrastructure.

Further evolution of supply chain management information systems strongly depends upon the success of the service-oriented architecture, which becomes more influential as more vendors provide functionality of their enterprise and supply chain applications as services.

The evolution of service-oriented architecture enables easier incorporation of decision-modeling components into the supply chain management information system. While many decision-modeling applications already provide an adequate technological support for integration with other parts of the supply chain information system, computational inefficiency remains an obstacle for several complex decision-making problems, including the supply chain configuration problem. From the information technology perspective, this issue is becoming less of a concern as technologies, such as grid-computing gain mainstream acceptance.

References

Agostinho C, Ducq Y, Zacharewicz G, Sarraipa J, Lampathaki F, Poler R, Jardim-Goncalves R (2015) Towards a sustainable interoperability in networked enterprise information systems: trends of knowledge and model-driven technology. Comput Ind, in press

Ahn HJ, Childerhouse P, Vossen G, Lee H (2012) Rethinking XML-enabled agile supply chains. Int J Inform Manag 32:17–23

Angles R (2005) RFID technologies: supply-chain applications and implementation issues. Inform Syst Manag 22:51–65

Bowersox DJ, Closs DJ, Cooper MB (2012) Supply chain logistics management. McGraw-Hill, Boston

Butner K (2010) The smarter supply chain of the future. Strat Leader 38:22–31

Dolk DR (2000) Integrated model management in the data warehouse era. Eur J Oper Res 122:199–218

Fleischmann B, Koberstein A (2015) Strategic network design. In: Stadtler H, Kilger C, Meyr H (eds) Supply chain management and advanced planning. Springer, Heidelberg, pp 107–123

Fuchs C, Otto A (2015) Value of IT in supply chain planning. J Enterprise Inform Manag 28:77–92

Helo P, Szekely B (2005) Logistics information systems: an analysis of software solutions for supply chain co-ordination. Ind Manag Data Syst 105:5–18

Hernández J, Mula J, Poler R (2014) Collaborative planning in multi-tier supply chains supported by a negotiation-based mechanism and multi-agent system. Group Decis Negot 23(2):235–269

Kwak K, Bae N, Cho Y (2014) Smart logistics service model based on context information. Lect Notes Electr Eng 301:669–676

Li Q, Zhou J, Peng QR, Li CQ, Wang C, Wu J, Shao BE (2010) Business processes oriented heterogeneous systems integration platform for networked enterprises. Comput Ind 61 (2):127–144

Lundell B, Lings B, Persson A, Mattsson A (2006) UML model interchange in heterogeneous tool environments: an analysis of adoptions of XMI 2. Lect Notes Comput Sci 4199:619–630

Mezgár I, Rauschecker U (2014) The challenge of networked enterprises for cloud computing interoperability. Comput Ind 65:657–674

Nurmilaakso J (2008) EDI, XML and e-business frameworks: a survey. Comput Ind 59 (4):370–379

Nyman HJ (2012) An exploratory study of supply chain management IT solutions. Proceedings of the annual Hawaii international conference on system sciences, pp 4747–4756

Qu T, Huang GQ, Zhang Y, Dai QY (2010) A generic analytical target cascading optimization system for decentralized supply chain configuration over supply chain grid. Int J Prod Econ 127:262–277

Radke AM, Tseng MM (2015) Design considerations for building distributed supply chain management systems based on cloud computing. J Manuf Sci Eng Trans 137:1–11

Siau K, Tian Y (2004) Supply chains integration: architecture and enabling technologies. J Comput Inform Syst 44:67–72

Stadtler H, Kliger C (2005) Collaboration planning. Supply chain management and advanced planning. Springer, New York

Zarghami A, Sapkota B, Eslami MZ, Van Sinderen M (2012) Decision as a service: separating decision-making from application process logic. Proceedings of the 2012 I.E. 16th international enterprise distributed object computing conference. EDOC, pp 103–112

Chapter 11
Data Integration Technologies

11.1 Introduction

The models presented in the previous chapters use knowledge of supply chain structure to represent the supply chain. Additionally, the parameters of the models were assumed as given and limited attention was devoted to estimation of these parameters. Data driven and statistical methods on the other hand can be used to uncover unknown structural relationships within the supply chain and provide methods for gathering and estimation of input data necessary for supply chain decision-making. Additionally, the data availability recently has increased dramatically making data driven approaches and attractive alternative for strategic supply chain analysis. That has opened a way to a range of new data gathering and supply chain analysis methods based on data integration from various sources. These methods follow a data driven approach implying that the primary means of analysis and decision-making are data processing operations. They are intricately intervened with technologies used for data integration and analysis.

Liu et al. (2014) point out that data integration is a significant challenge in the supply chain environment. They used an integrated data pipeline for recording transactions among the supply chain partners and the ETL process is used to load data in a common data warehouse. Business intelligence methods are used for data analysis supported by a common managerial model. Data integration allows for evidence based supply chain risk assessment. Hahn and Packowski (2015) suggest application of in-memory analytics for processing large data volumes in supply chains and also develop a framework for analytics applications in supply chain management. The framework shows application of data-driven approaches in data exploration and monitoring approaches. In line with latest developments in computing, Neaga et al. (2015) propose using a cloud based platform for big data analytics supporting logistics service. The key principles of the platform design are data integration to support various types of information and end-users, usage of

C. Chandra, J. Grabis, *Supply Chain Configuration*,
DOI 10.1007/978-1-4939-3557-4_11

open data integration standards, data model unification, and collaborative utilization of analytical services.

Section 11.2 discusses importance of the data driven approach in the overall supply chain configuration framework. Section 11.3 provides overview of the approach subsequently elaborated in Sects. 11.4 and 11.5. Application of the approach for multi-objective facility location is demonstrated in Sect. 11.6.

11.2   Purpose

Data integration has a number of applications in the supply chain configuration methodology. It is used for processing of both input and data of supply chain configuration initiative. Its primary purpose is providing data to decision-making models. Additionally, some of the supply chain configuration decisions can be made directly according to the results of data exploration and analysis, Regardless of the way alternative supply chain configurations are created, data integration technologies are used for presentation of the modeling results to facilitate decision-making concerning the final supply chain configuration to be implemented. Once the supply chain configuration is implemented, data integration technologies are used to monitor supply chain performance.

Centralized approaches to data integration such as data warehousing and spreadsheets have long dominated (Dolk 2000). Currently, on-demand and distributed methods are becoming more widely used. The purpose of these methods is to:

- Provide more flexibility for data processing.
- Get access to the most current data.
- Expand the range of data used for decision-making.

Traditional data warehouses operate on a premise that decision-making data needs are known in advance and that these data can be gathered and structured in advance. However, that is not always the case, especially for supply chain reconfiguration during the execution phase. For instance, suppliers in Southeast Asia are characterized by different attributes than substitute suppliers in South America. The on-demand approach provides data as necessary for a particular decision-making situation and is more suitable for agile and reconfigurable supply chains.

Traditional modeling methods such as mathematical programming also have a rigid structure from the data processing perspective (i.e., the models and often model-solving algorithms, too, are designed for a specific set of variables and parameters). The data analysis based approaches are more flexible and allow for rapid incorporation of new data items in decision-making, as necessary.

There is a wide range of data integration and analysis methods, which can be tailored for data driven supply chain configuration. This chapter introduces a method focusing on data gathering from web services and ranking based supply chain facilities location.

11.3 Approach Overview

The data driven approach to supply chain configuration is part of the overall supply chain configuration framework and decision support systems. Within this framework, it emphasizes data gathering and representation of modeling results as well as uses a specific data intensive supply chain configuration model. These key elements of the data driven approach are shown in Fig. 11.1

The conceptual model and selection of the supply chain configuration methods made in the decision support systems determines modeling input data requirements. The data gathering module is responsible for obtaining these data from external data and to format them according to needs of the model solving algorithm. The predefined modeling input data or intermediate data necessary for data gathering are data readily available. Spatial data and non-spatial data catalogue services are used to discover appropriate external data sources for necessary spatial and non-spatial data, respectively. Services catalogues like Universal Description, Discovery and Integration (UDDI) or OpenGIS Catalogue Services can be searched for suitable services. In cases when there are multiple suitable services, the best service is selected using the Service Selection component by analyzing Quality of Service (QoS) data from QoS data repositories (see Luo et al. (2004) on selecting spatial services and Buccafurri et al. (2008) on selecting non-spatial services). QoS measures indicate data access speed and service reliability. Service selection procedures also account for issues related to data compliance with requirements and data quality. Service selection and QoS evaluation are essential because distributed data are owned by third parties and accessed over network, and data and their access quality cannot be assured without prior evaluation.

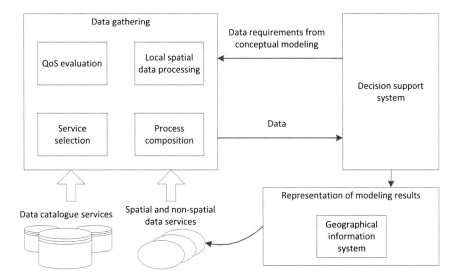

Fig. 11.1 The key elements of the data driven approach

In some cases, it is more efficient to perform spatial data processing tasks locally rather than requesting processing operations from remote service providers. The Local Spatial Data Processing component provides such functions as some geocoding functions, geographic coordinate conversion, and calculation of distances between pairs of points with known coordinates. The Process Composition component is used to define a sequence of data retrieval and processing operations. These operations frequently depend upon each other and, therefore, the order of their involvement must be specified. The data retrieval process can be defined using general purpose programming languages or process composition languages WS-BPEL (Web Service Business Process Execution Language). The data gathering result is a set of modeling data presented in the format required by the model-solving algorithm.

The model-solving algorithm is responsible for solving the facility location model using the provided input data. The geographical information system is used to represent the modeling results. The representation is created by combining spatial data layers provided by different external data sources.

The data intensive supply chain configuration model considered in this chapter deals with facility location in one of the supply chain tiers. The distinctive feature of this model is that it is able to take into account a large variety of decision-making factors (Grabis et al. 2012). The facility location problem is defined as a multi-objective mixed-integer programming model. The model selects potential facilities from a set of predefined alternative locations by maximizing an aggregated facility goodness indicator as a weighted sum of several facility goodness indicators corresponding to criteria describing potential locations. Different sets of the facility goodness indicators can be used. Indicators characterizing number of customers, number of competitors, and real-estate cost are used in this chapter. These correspond to the size of market, the location of key competitors and the cost of land factors deemed as important facility location factors by Bhatnagar and Sohal (2005). The data driven approach implies that these factors can be changed dynamically without affecting the overall structure of the model.

11.4   Data Gathering

Data characterizing all decision-making factors should be gathered. Spatial and non-spatial data catalogues are used to identify potential data sources satisfying modeling input data requirements. Identified data sources and their characteristics are given in Table 11.1. The Geocoding service is able to return results in multiple formats though CSV (Comma Separated Values) is used as the result contains only two elements—longitude and latitude. The Competitor data service returns results in two formats, from which KML was chosen. Each request to this service returns multiple KML documents with limited number of records in each document. The Real estate service returns median price per square feet based on city name or zip code. The Population service returns population data in the XML format. The

Table 11.1 List of external data sources

#	Data source	Function	Data format	Interface
1	Geocoding service	Converting addresses of facility locations into geographical coordinates	CSV, XML, KML, JSON	Web service (WS)
2	Competitors service	Finding spatial location of businesses of specified type	KLM, JSON	WS
3	Real estate service	Finding real estate data for a specified location	XML	REST (representational state transfer) style WS
4	Population service	Finding number of customers in a specified area	GML	WFS

results can be filtered by bounding box. Unfortunately, the service is unable to filter results based on radius. To overcome this problem, coordinates of a bounding box that contains a circle with the specified radius are calculated and the Population service is queried using these coordinates. After receiving the results from the Population service, the Local spatial data processing component filters only those points that fall inside the circle.

11.5 Process Composition

Data from the external data sources are retrieved and formatted in the specific order, which is established using the Process composition component. The data gathering process is described in Fig. 11.2. The UML swimlanes are used to show components/services responsible for performing a particular data gathering activity. The diagram also shows interdependencies among the data gathering activities and parallelization opportunities.

Addresses of the potential facility locations are provided as the predefined modeling input data. The Local spatial data processing component orchestrates the whole data gathering process and performs some of the data processing activities. In order to expedite data gathering, the activities are parallelized, if possible taking into account data interdependencies. The Geocoding service is used to determine coordinates of the potential facility locations according to their address. Those coordinates are used to query the Competitor service to retrieve information about the nearby competitors and to query the Population service to retrieve the number of customer in the proximity of potential locations. The Real estate service, which is a non-spatial data service, is queried by providing city name and zip code of the potential facility location. The diagram shows that the Competitors service and the Population service can be invoked only after the geocoding because of the data interdependencies, while the real estate data can be retrieved in parallel to the geocoding. The straight-line distance calculation between the potential locations using their coordinates is performed at the Local spatial data processing component.

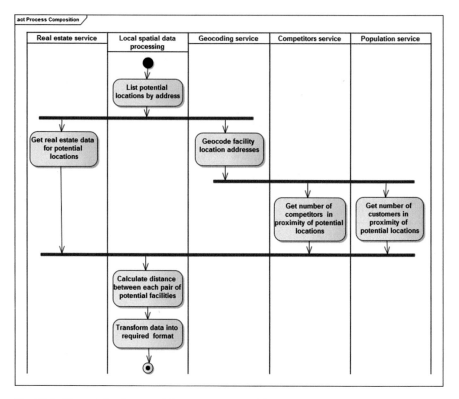

Fig. 11.2 The activity diagram of the process composition

This activity also could be performed using an external service. However, the local processing is computationally more efficient in this case. The data transformation according to requirements of the model-solving algorithm is the last step of data gathering process. In this case, data are passed to the model-solving algorithm as pointers to data arrays in computer memory.

This activity diagram is used to implement the data gathering process. Calls to the external services are implemented so that the data sources can be replaced with other data sources without affecting the overall process composition, and different types of external services can be used. The data are accumulated in an XML format data file, and data mappings are used to transform data to and from formats and data structures supported by the external services. More technical details on the implementation of the data gathering process can be found in Kampars and Grabis (2011).

11.5.1   *Selection Model*

A multi-objective model is formulated to locate facilities according to the number of customers, number of competitors, and real-estate cost criteria. Each potential site is characterized by its customer, competitor, and land cost indicators. The customer indicator characterizes the number of customers in proximity of the site. The competitors indicator characterizes the number of competitors in the proximity of the site. The land cost indicator represents the land purchasing price in the area the site belongs to. Data gathered according to the aforementioned data gathering process will be used to solve the model.

Notation

i	Subscript indicating a potential site
X_i	A binary potential facility site open indicator, $X_i \in \{0, 1\}$
a_i	Customer indicator at site i
c_i	Competitor indicator at site i
l_i	Land cost indicator at site i
r	Coverage radius
B_i^s	Set of points j falling within r around i, where s indicates a type of point
α_j	Number of customers at point j
d_{ij}	Distance between site i and point j
$\Delta_{ii'}$	Distance between two potential facilities i and i'
$v_{ij} = \begin{cases} \exp(-u_{ij}), u \le 1 \\ 0, u_{ij} > 1 \end{cases}$	Weight coefficient, where $u_{ij} = \frac{d_{ij}}{r}$
N	Number of potential sites
P	Maximum number of sites
H	Number of site selection criteria
Z	Aggregated facility goodness indicator
Z_h	Facility goodness indicator for selection criterion h, $h = 1, .., H$
w_h	Weight coefficient characterizing importance of each selection criterion, $h, h = 1, .., H$

11.5.1.1   Model Formulation

The multi-objective function maximizes the aggregated facility goodness indicator

$$Z = \max \sum_{h=1}^{H} w_h Z_h.$$

In the general case, Z_h can be expressed as the sum of indicator values at each selected facility location. If the number of customers, number of competitors, and real-estate cost are used as the selection criteria, then Z_1, Z_2 and Z_3 are computed using expressions (11.1), (11.2), and (11.3), respectively. The customer indicator Z_1 is computed as

$$Z_1 = \sum_{i=1}^{N} a_i X_i, \qquad (11.1)$$

where $a_i = \sum_{j \in B_i^1} v_{ij} \alpha_j$ is a sum of customers in proximity of potential location i exponentially weighted by the distance. This indicator is used as a proxy for customer demand. In order to calculate the indicator value, a_i is scaled to range between 0 and 1.

The competitor indicator Z_2 is computed as

$$Z_2 = \sum_{i=1}^{N} c_i X_i, \qquad (11.2)$$

where $c_i = \sum_{j \in B_i^2} v_{ij}$ is a sum of competitors in proximity of potential location i exponentially weighted by the distance. In order to calculate the indicator value, c_i is scaled to range between 0 and 1.

The real estate cost indicator Z_3 is computed as

$$Z_3 = \sum_{i=1}^{N} l_i X_i, \qquad (11.3)$$

where l_i is scaled to range between 0 and 1.

Maximization is performed subject to the following constraints. Constraint (11.4) restricts the number of facilities to be selected:

$$\sum_{i=1}^{N} Y_i \leq P. \qquad (11.4)$$

Constraint (11.5a) implies that the distance between two open facilities should be larger than $2r$:

$$\Delta_{ij} X_i X_j \geq 2r, \forall i, j \qquad (11.5a)$$

This constraint is similar to constraints often used in traditional facility location models assigning each demand point exclusively to a single facility. In order to avoid nonlinearity, multiplication $X_i X_j$ is replaced by the following constraints:

$$\Delta_{ij} Y_{ij} \geq 2r Y_{ij}, \forall i, j, \qquad (11.5b)$$
$$Y_{ij} \leq X_i, \forall i, j \qquad (11.5c)$$
$$Y_{ij} \leq X_j, \forall i, j \qquad (11.5d)$$

$$Y_{ij} \geq X_i + X_j - 1, \forall i, j, \qquad (11.5e)$$

where Y_{ij} is a binary variable.

Additionally, constraint (11.6) is introduced to impose symmetry:

$$Y_{ij} = Y_{ji}, \forall i, j. \qquad (11.6)$$

In order to compute the aggregated location goodness indicator, a weighted sum of individual facility goodness indicators is computed. Importance of each indicator can be determined according to the results of empirical studies on practical importance of different facility location criteria. The model presented above uses just three facility location criteria although other criteria could be readily incorporated, if necessary.

The data sources identified in Sect. 11.4 are used to assign values to the model's parameters. The first data source is used for intermediate processing and for calculating $\Delta_{ii'}$. The second source is used to calculate c_i. The third data source is used to calculate l_i. The fourth data source is used to calculate a_i.

11.6 Experimental Studies

The proposed approach is evaluated in the experimental studies. The experimental studies explore both data gathering and model-solving aspects.

11.6.1 Problem Description

The sample facility location problem considered in this paper deals with locating fast food restaurants. There are a number of preselected potential facility location sites and the total number of facilities to be open is limited. It is aimed to locate restaurants at sites having the largest number of customers and the smallest number of competitors in its proximity and having acceptable real estate costs.

The only predefined modeling input data necessary are addresses of potential facility location sites. A thousand addresses are randomly generated using the address listings in Michigan's Yellow Pages. The number of potential sites N is varied from 20 to 200 corresponding to small to medium size facility location problems. The number of facilities to be open is varied from $N/10$ to $N/5$. The values of the coverage radius are 5, 10 and 15 km. The model is solved using: (1) simple rating (denoted by SR); and (2) direct solving using a commercially available mathematical programming package (DS). The simple rating is used as a benchmark to evaluate the efficiency of using mathematical programming. The importance of each factor in the objective function is set according to the survey results by Bhatnagar and Sohal (2005). Only w_1 is increased threefold to provide incentives for opening larger number of facilities. That gives $w_1 = 72.4$, $w_2 = -6.9$ and $w_3 = -21.4$ (the negative values are used to indicate that these factors discourage opening new facilities).

11.6.2   Data Gathering and QoS Evaluation

Computational experiments of data retrieval are conducted after the data retrieval process has been set-up as described in Sect. 11.5. The cumulative data retrieval time depends on N. In order to evaluate data retrieval time, data are sequentially requested for 1000 potential facility locations. The cumulative data retrieval time according to the number of requests made for each data source is given in Fig. 11.3. The Geocoding service is the fastest while the Population data service is the slowest. It should be noted that the Population data service returns a large data set for each request (size of the data set varies) while the Geocoding service returns a few data items. The slope of data retrieval time curve is larger than one for the Competitors and Population services because the cumulative data time is affected by size of the return data set and QoS issues. The QoS issues in the form of sudden spikes in data retrieval time most profoundly can be observed for the Population data service.

In order to gain insights in QoS characteristics of each service, more detailed studies on QoS evaluation have been conducted. QoS is measured by the response time (i.e., the average time in which service responded to requests made by DA) and the percentage of failed request. The request is qualified as a failed request if no

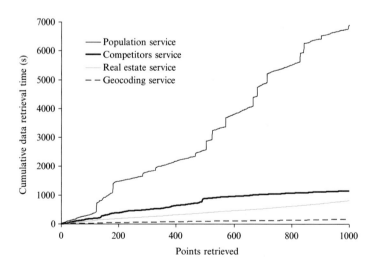

Fig. 11.3 Data retrieval time

Table 11.2 QoS measures

Measure	Geocoding service	Competitor service	Real estate service	Population service
Failed requests (%)	0	15.1	11.9	46.5
Average response time (seconds)	0.16	1.14	0.81	6.89

response or empty data was returned. The QoS measures are evaluated over the period of 24 h, and the results are summarized in Table 11.2.

These results show that QoS varies considerably. The fastest is the Geocoding service. It also has no failed requests and is able to geocode all of the addresses tested. The Real estate service has fast response time while there is a large number of failed requests. The failures were caused by missing data. The average real estate price was used as a substitute. For the Competitor service, request failures were caused both by missing data and network related errors. In the former case, potential facility locations were excluded from further modeling, and requests were retried in the latter case. The most unstable was the Population data service. The response time of this service was long, and it returned results only in 46.5 % of all cases on the first attempt. This service also was the only service, which was not available for prolonged periods of times. The total data gathering time for 200 alternative locations is about 30 min. From the decision-modeling perspective, missing data and low availability are the most undesirable characteristics of data sources.

The response time for the Competitor data service and the Population data service substantially depends on the coverage radius r. For instance, The Competitor data service has the average response time of 0.76 s for $r = 5$ and 1.9, for $r = 15$.

11.6.3 Facility Location Results

Using the gathered data, the facility location problem is solved using a commercially available solver. The facility location results are evaluated by comparing facility location using simple ranking (SR) (i.e., constraint (11.5a) is not taken into account) and facility location maximizing the aggregated facility goodness indicator using mathematical programming (DS). Several experimental cases are constructed by varying the coverage radius, the number of potential sites and the maximum number of open facilities. For each experimental cell, the model-solving is performed for ten different randomly generated sets of candidate facility locations (the sets are drawn from the initial list of 1000 addresses).

Table 11.3 reports the ratio between the aggregated facility goodness indicators obtained using SR and DS. The results are the same regardless of the

Table 11.3 Comparison of model solving algorithms

N	P	$r = 5$ Z(SR)/Z(DS)	$r = 10$ Z(SR)/Z(DS)	$r = 15$ Z(SR)/Z(DS)
20	2	1	1	0.99
	4	1	1	0.98
50	5	1	0.96	0.94
	10	0.99	0.95	0.94
100	10	0.97	0.93	0.90
	20	0.96	0.93	0.90
200	20	0.93	0.91	0.94
	40	0.94	0.91	0.93

Z(SR)—denotes the aggregated facility goodness indicator obtained using SR
Z(DS)—denotes the aggregated facility goodness indicator obtained using DS with $T_{DS} = 5\,h$

algorithm used for small values of N and r because there are a few nearby locations and constraint (11.5a) has limited impact on the solution. For the upper level values of N and r, the simple ranking yields up to 10 % (averaged over ten sets of candidate locations) worse result than the optimal result. The number of selected units usually is only a fraction of P due to the impact of competition and real estate cost limiting incentives to open maximum number of facilities. For the upper level values of r, there is also a limited number of candidate locations satisfying constraint (11.5a).

11.6.4   Representation of Results

The important feature of spatial data processing is the ability to represent data graphically using various cartographical tools. Figure 11.4 shows a sample facility location results for one of the cases with $N = 20$, $P = 4$ and $r = 15$ (not all units are visible). The picture is obtained by applying multiple layers of spatial data on each other. Initially, the population density and competitors' location data are retrieved from data sources and shown in the map (the land cost is a scalar data type and is not represented on the map). This is followed by indicating locations of the potential facilities. After modeling has been completed, the final layer containing selected facilities is applied and the modeling results are represented on the map. The figure shows that there is strong preference for locations with high population density. Technically, the graphical representation is created using the KLM format.

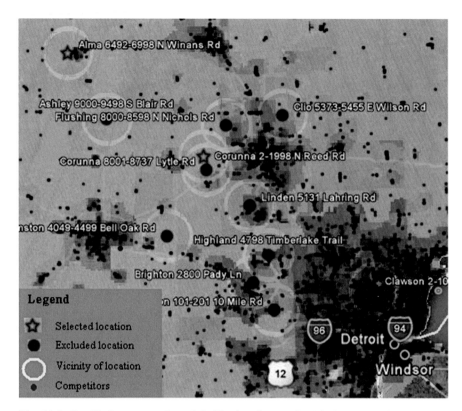

Fig. 11.4 Graphical representation of facility location results (*shades* are used to represent population density) (Grabis et al. 2012)

11.7 Summary

The data driven approach capitalizes on increasing availability of data for supply chain management and decision-making purposes. Its main advantage is increased modeling and decision-making flexibility as new data can be gathered and taken into account dynamically which is of particular importance in the case of agile and reconfigurable supply chains.

There are also multiple challenges to be considered for successful utilization of distributed spatial data. Availability of public data sources is limited while maintaining relevant data at the enterprise is challenging mainly due to expensive data updating and complex infrastructural requirements. There is a large number of standards and technologies used in distributed data processing and the data retrieval process consists of multiple interrelated steps. Standardized interfaces and data formats are used to avoid incurring significant data integration overhead in the case of on-demand data gathering. Data quality and data retrieval characteristics vary significantly for different data sources and data gathering time can be long, especially, for high resolution data or data aggregated over large geographic areas. The data retrieval process automation allows addressing QoS problems by re-querying data sources. Data from external sources can be retrieved in the XML format, which can be processed with ease. However, data providers use different XML based standards (e.g., GML and KML) which requires an extra effort to attain understanding of the data structure. Additionally, many data sources are poorly documented and providers occasionally modify the data structure without notification. Cartographical resources enable representation of modeling results through a combination of data representation layers from multiple sources involving manual operations.

The proposed facility location model can be expanded to include additional criteria subject to data availability. While majority of existing facility location models focus on optimizing facility location costs or travel time related measures, the proposed model attempts to locate facilities according to a wide range of contextual characteristics.

References

Bhatnagar R, Sohal AS (2005) Supply chain competitiveness: measuring the impact of location factors, uncertainty and manufacturing practices. Technovation 25(5):443–456

Buccafurri CF, De Meoc P, Fugini M, Furnari R, Goy A, Lax G, Lops P, Modafferi S, Pernici B, Redavid D, Semeraro G, Ursino D (2008) Analysis of QoS in cooperative services for real time applications. Data Knowl Eng 67(3):463–484

Dolk DR (2000) Integrated model management in the data warehouse era. Eur J Oper Res 122:199–218

Grabis J, Chandra C, Kampars J (2012) Use of distributed data sources in facility location. Comput Ind Eng 63(4):855–863

Hahn GJ, Packowski J (2015) A perspective on applications of in-memory analytics in supply chain management. Decis Support Syst 53:591–598

Kampars J, Grabis J (2011) An approach to parallelization of remote data integration tasks. Sci Proc RTU Computer Sci 49:24–30

Liu L, Daniels H, Hofman W (2014) Business intelligence for improving supply chain risk management. Lect Notes Bus Inform Process 190:190–205

Luo Y, Liu X, Wang W, Wang X, Xu Z (2004) QoS analysis on web service based spatial integration. Springer, Berlin/Heidelberg, pp 42–49

Neaga I, Liu S, Xu L, Chen H, Hao Y (2015) Cloud enabled big data business platform for logistics services: a research and development agenda. Lect Notes Bus Inform Process 216:22–33

Chapter 12
Mobile and Cloud Based Technologies

12.1 Introduction

Enterprise applications, advanced planning systems, and enterprise application integration technologies provide a well-established way of providing information technology support for supply chain management. Despite enormous gains in flexibility of these technologies their general characteristics remain corresponding to the lean and flexible supply chain strategies. These technologies are based on standardization and require relatively large up-front investments and setup time, thus limiting supply chain reconfiguration opportunities.

Recently, a number of new trends have emerged in information technology. These include smart systems, proliferation of mobile technologies and cloud computing. The smart systems are based on integration of sensor systems, processing of large data volumes and software to optimally control different kind of physical and computational systems. The typical examples of the smart systems are smart homes and sensor-based logistics systems. For instance, Karakostas et al. (2012) describe usage of sensors and intelligent decision-making systems in transportation of short-life cycle products. The mobile technologies allow capturing data, transaction processing and decision-making outside the traditional office environment. That increases the speed of information flows and more importantly give rise to new types of business processes. For instance, in the shipping industry, data about cargo location are updated only after a vessel has arrived in the harbor while mobile applications such as Ship Finder[1] allow for tracking the vessels in real time. The mobile technologies together with the sensing technologies are also major enablers of context awareness (Hong et al. 2009), making supply chain agility possible. The cloud computing is providing availability of scalable information processing services over the Internet in the on-demand mode. It is a major enabler of seamless

---

[1] http://shipfinder.co

© Springer Science+Business Media New York 2016
C. Chandra, J. Grabis, *Supply Chain Configuration*,
DOI 10.1007/978-1-4939-3557-4_12

information sharing, regardless of the physical location and offers tremendous computational resources for decision-making.

This chapter investigates opportunities of using these technologies in supply chain management and their impact on supply chain configuration in particular. It is argued that the mobile and cloud computing technologies are giving rise to a new type of supply chain where physical and digital flows are fused together. To illustrate a cloud chain, an example of e-retailing supply chain is used. Many value added and critical supply chain functions in this supply chain are performed in the cloud environment. A model configuration of the cloud chain is elaborated, and it shows interdependencies between selection of physical and digital supply chain units.

12.2 Purpose

From the information technology perspective, reconfigurable supply chain should possess ability of:

- Quickly gathering information about the current status of the supply chain and to store historical data
- Sharing information among supply chain units and supporting collaborative business processes
- Performing computationally demanding evaluation of alternative supply chain configurations
- Quickly integrating new supply chain units or functions into the supply chain information systems
- Dealing with heterogeneity and providing customized solutions in the global supply chain.

The emerging information technologies allow dealing with these challenges. The sensing technologies provide means for real time information gathering. For instance, sensors are used to constantly monitor condition of perishable goods during the transportation process and the transportation mode can be changed, if necessary (Metzger et al. 2012). The mobile technologies are used to quickly transmit real-time information and to enable communications among supply chain partners. The cloud computing is vital for storing the sensing data and for providing enough computational power for processing these data. It also eliminates the need for building up expensive supply chain infrastructure. The standardized web services available over the cloud allow for quick integration of supply chain partners and reduce the impact of heterogeneity on complexity of supply chain integration.

Klein and Rai (2009) suggest that a strategic approach is needed to information integration in supply chains, and Swaminathan and Tayur (2003) identifies opportunities for using emerging information technologies in e-business supply chain. One of the options for improving information flows is improvements in information logistics. It is also confirmed by findings that information accuracy and relevance is among the key factors affecting web-site quality in e-business (Hernandez et al. 2009). A hub based approach can be used in integrating the physical and information flows in supply chains (Trappey et al. 2007). However, in the case of highly distributed and heterogeneous supply chains as in e-retailing, a service oriented approach could be a more attractive option (Candido et al. 2009).

12.3 Cloud Chain

Utilization of mobile and cloud based technologies allow for a new kind of supply chain referred here as cloud chains or digital supply chain.[2] A cloud chain is the supply chain where a large share of value is added through using virtual supply chain units or communication channels and products/services delivered are made possible by using cloud based and mobile technologies. These products and services are not merely alteration of existing products and services to new technologies but new products and services not previously available. A distinction between virtual and physical supply chain units is made. Three types of units are present in cloud chains:

- Physical supply chain units—traditional supply chain units located at certain physical locations and processing physical materials and products
- Virtual supply chain units—units that are not dedicated to certain physical locations providing digital products or providing digital value added or critical supply chain services
- Information processing service units—units providing supporting information processing services.

The distinction between virtual production units and information processing units is made according to their contribution to creation of the cloud chain's end product or service. The virtual productions units perform primary value-added or critical activities while the information processing units perform supporting activities. The critical activities are those without which the very existence of the supply chain end-product or service would be impossible.

Figure 12.1 shows an illustrative cloud chain for manufacturing of mobile phones (only a few representative units are shown). The cloud chain is represented in three layers each containing the corresponding type of units. The circuits and screen suppliers are traditional physical supply chain units providing materials for the assembly. However, digital printing recently has emerged as a promising technique for attaching antennas to the mobile phone.[3] The digital printing is

---

[2] http://insights-on-business.com/electronics/3d-printing-transforming-the-supply-chain-part-1/

[3] http://www.engineering.com/3DPrinting/3DPrintingArticles/ArticleID/5800/3D-Printed-Antennas-Could-Reduce-Size-Cost-of-Mobile-Devices.aspx

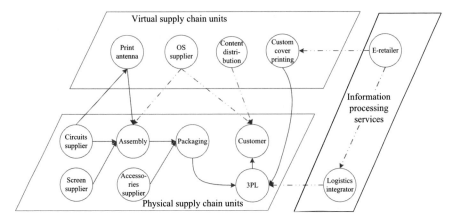

Fig. 12.1 A sample cloud chain

perceived as a virtual supply chain unit since it receives supplies both physically and electronically (i.e., antenna's design) and in the future digital printers could be available at many different locations, making them virtually location independent. The phone Operating Systems (OS) is completely digital product, which is delivered to customers electronically (denoted by the dash-dot arrow). Similarly, the digital content such as applications, audio and video streams are also electronic products delivered directly to the customer, and they are an essential part of the phone ecosystem (Basole and Karla 2011). The customer buys the phone from the e-retailer. The e-retailer also provides access to custom phone cover digital printing services.[4] It is represented as an information processing unit because it is deemed as a supporting service. Finally, the phone physically is delivered to the customer by the 3PL provider, which relies on supporting services provided by the Internet based logistics integrator. The logistics integrator provides payment, tax and duties, order tracking, and other services associated with global product delivery.

The supply chain unit selection in cloud chains is affected by a number of different factors than used in traditional supply chain configuration. These factors are related to the Quality of Service (QoS) criteria used in web service selection (Strunk 2010). They include service response time, service availability and service reliability. It is expected that physical supply chain configuration decisions and virtual/information processing supply chain configuration decisions are mutually interdependent. Therefore, the cloud chain configuration includes joint selection of all kinds of supply chains units according to their respective selection criteria.

---

[4] http://www.wired.com/2013/01/nokia-3d-print-case/

12.4   Web Service Selection Model

A model for joint design of the physical and information flows is developed for supply chains, where a significant part of supply chain activities take place in an electronic form. Supply chains by e-retailers such as Amazon.com, Macy's[5] belong to this group of supply chains. The physical flow represents the flow of products from supplier's to e-retailer's facilities and final delivery of products to end-customers is usually done by a 3PL provider. The information flow represents different on-line services to customers and supply chain partners. These services include product information services, payment services, insurance services, shipment tracking services, and others. The services can be provided by the same partners providing the physical processing or by partners specializing in delivering electronic information processing services. For instance, Borderfree[6] acts as an integrator for e-retailers providing end-to-end information processing services.

12.4.1   Physical and Internal Information Flow

A business process model is used to represent the physical and information flows and their processing in supply chains. It is assumed that the physical data flow and supply chain units mainly dealing with processing of physical products are represented as a single entity while supply chain units mainly dealing with information processing are represented as independent units. Therefore, the physical supply chain units are represented in the business process as lanes in a single pool (Fig. 12.2), and the electronic supply chain units are represented as separate pools (see Sect. 12.4.2).

The physical flow of products is initiated by detecting the customer demand without specifying how the demand is detected. Suppliers are responsible for supplying the products. The e-retailer is the focal unit in the e-retailing supply chain and its main task is to sell products to customers. The e-retailer can also operate storage and distribution facilities. The 3PL providers are responsible for delivery of products to customers. The internal information flow accompanies the physical flow of products. It is referred to as the internal information flow because information processing is perceived as an essential part of the physical products' processing tasks. Data objects are used to represent internal supply chain information flows. Only the main data objects such as sales order, purchasing requisition, delivery note, and delivery confirmation are referenced in the model. This supply chain representation does not include the reverse supply chain flow for simplicity.

An integrated physical and information flow model is created in order to capture interrelationship among the physical and electronic supply chain units.

---

[5] http://www.amazon.com, http://www.macys.com

[6] http://www.borderfree.com

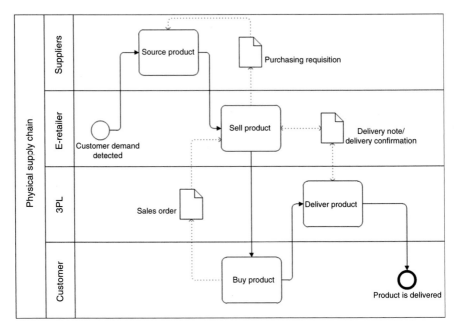

Fig. 12.2 The physical flow in supply chains

The electronic supply chain units are represented each in a separate pool named as a service unit with a specific type. These pools represent abstract service providers. The actual service providers can provide several of the services required, some of them act as service aggregators and some services can be provided by the physical supply chain units. The interrelationships are shown as message flows among the pools. The message flow shows only purely electronic information processing activities. For instance, a shipment activity includes shipment data processing, shipment confirmation and other operations but these information processing activities are perceived as an essential part of physical activities and are included in the **Deliver product** task.

The model defines main types of the electronic service units present in the e-retailing supply chains in the global setting. These types include:

- Product information services—detailed information possibly aggregated from multiple sources is provided about each product offered by the e-retailer
- Import/export services—checks on import and export restrictions from one country to another for certain products, i.e., the service rejects selling a product in certain countries where specific licensing rules are applicable
- Customs and taxes services—calculation of appropriate taxes depending upon the customer location is performed
- Payment services—multi-currency processing of payments using different payment channels is performed and restrictions concerning availability of the

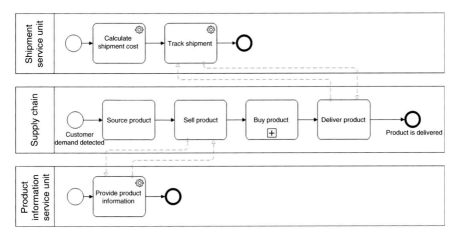

Fig. 12.3 The message flow between the supply chain and the product information and shipment services

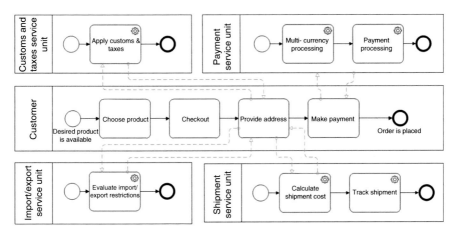

Fig. 12.4 The message flow for the **Buy product** task

payment channels are applied, e.g., credit cards only from specified countries are accepted

- Shipment services—if multiple shipment modes are available, the most appropriate alternative with regard to the destination and delivery time is determined and shipment tracking is provided independently of the 3PL provider; especially, if multiple logistics providers are used for delivery.

The list of service types is not exhaustive and other types of services can be used such as fraud detection and shipment insurance. Figure 12.3 shows the physical e-retailing supply chain process along with the necessary electronic services. The expansion of the **Buy product** task is given in Fig. 12.4. The message flow for this task is shown only at the sub-process level. The product information service provides information to the **Sell product** task and is responsible for providing as rich information about the product as possible. The shipment service is invoked during the product delivery to provide opportunities for tracking the product delivery.

Majority of message flows are associated with the **Buy product** task. Services are invoked to provide an accurate estimate of the total ordering costs for the customer. For local e-retailers, this operation usually is straightforward but much more comprehensive information should be gathered for global e-retailers. The message flows should ensure information about applicable taxes, import/export restrictions, delivery options, and international payment processing. This information is specific to the customer location.

In the given process model, it is assumed that the e-retailer manages both the physical and electronic sales process by itself. Another possibility is that a traditional retailer deals only with the physical sales while the electronic part is provided by a sales service provider.

12.4.2   Model

The supply chain business process models show interactions among the physical and electronic supply chain units. The supply chain configuration problem is to select suitable physical and electronic supply chain units to optimize supply chain performance. In the case of e-retailing supply chain, products' suppliers, third party logistic provider, and web services for information processing are selected. The supply chain performance is measured by supply chain profitability and customer satisfaction affected by efficiency of information processing. The profitability is calculated as revenues from product sales after deducting sales expenses minus sourcing, delivery, and unit setup costs. The information processing efficiency is calculated as a weighted sum of web service QoS criteria, namely, response time, error rate, and reliability, which are among the most frequently used QoS.

The mathematical formulation of the model consists of the objective function (Eq. 12.1) and constraints (Eq. 12.7–12.13). Equations (12.2–12.6) are auxiliary measures used to the elements of the objective function. The notations used are defined in Table 12.1. The weights w_1 and w_2 are used to combine the physical units selection and web service selection criteria in a single objective function. The physical units selection is performed to maximize e-retailer's profit calculated as a difference between revenues R and sourcing cost C_1, delivery cost C_2 and fixed cost C_3. The web service selection is performed to maximize infrastructure processing efficiency L.

Equation 12.6 evaluates the information processing efficiency for the selected web services. The importance of each QoS criterion is determined by the weight factor v_l. Eq. 12.5 evaluates the fixed cost incurred by incorporating physical or electronic units in the supply chain.

Table 12.1 Notation

Notation	Description
N_p	Number or products
N_c	Number of countries where customer are located
N_v	Number of potential suppliers
N_l	Number of potential 3PL providers
N_s	Number of potential services
N_f	Number of required functions
$X_i \in \{0,1\}$	A decision variable indicating whether the ith supplier is selected or not
$Y_i \in \{0,1\}$	A decision variable indicating whether the ith service is selected or not
$Z_i \in \{0,1\}$	A decision variable indicating whether the ith 3PL provider is selected or not
S_{ij}	A decision variable determining the quantity of the ith product sold to customer in the jth country
Q_{ij}	A decision variable determining the quantity of the ith product sourced from the jth supplier
U_{ijk}	A decision variable determining the quantity of the ith product delivered by the jth 3PL provider to the kth country
σ_{ij}	Revenues from each item of the ith product sold in the jth country
π_{ij}	Purchasing prices of the ith product from the jth supplier
δ_{ijk}	Delivery cost for the ith product by the jth 3PL provide to the kth country
λ_i^1	The setup cost for the ith supplier
λ_i^2	The setup cost for the ith service
λ_i^3	The setup cost for the ith 3PL provider
β_{ij}	The value of the jth QoS attribute for the ith service
d_{ij}	Demand for the ith product in the jth country
γ_{ij}	Equals to one if the ith service supports the jth function and zero if not
τ_{ij}	Equals to one if the ith service is available in the jth country and zero if not
M	A large number

$$P(\mathbf{X}, \mathbf{Y}, \mathbf{Z}) = w_1(R - C_1 - C_2 - C_3) + w_2 L \rightarrow \max \qquad (12.1)$$

$$R = \sum_{i=1}^{N_p} \sum_{j=1}^{N_c} \sigma_{ij} S_{ij} \qquad (12.2)$$

$$C_1 = \sum_{i=1}^{N_p} \sum_{j=1}^{N_v} \pi_{ij} Q_{ij} \qquad (12.3)$$

$$C_2 = \sum_{i=1}^{N_p} \sum_{j=1}^{N_l} \sum_{k=1}^{N_c} \delta_{ijk} U_{ijk} \qquad (12.4)$$

$$C_3 = \sum_{i=1}^{N_v} \lambda_i^1 X_i + \sum_{i=1}^{N_s} \lambda_i^2 Y_i + \sum_{i=1}^{N_l} \lambda_i^3 Z_i \tag{12.5}$$

$$L = \sum_{i=1}^{N_s} \sum_{j=1}^{3} v_j \beta_{ij} Y_i \tag{12.6}$$

$$S_{ij} \leq d_{ij}, i = 1, \ldots, N_p, j = 1, \ldots, N_c \tag{12.7}$$

$$\sum_{j=1}^{N_l} U_{ijk} \leq S_{ik}, i = 1, \ldots, N_p, k = 1, \ldots, N_c \tag{12.8}$$

$$\sum_{j=1}^{N_c} S_{ij} \leq \sum_{k=1}^{N_v} Q_{ik}, i = 1, \ldots, N_p \tag{12.9}$$

$$\sum_{i=1}^{N_s} \gamma_{ij} Y_i = 1, j = 1, \ldots, N_f \tag{12.10}$$

$$\sum_{l=1}^{N_p} S_{li} \leq \sum_{k=1}^{N_s} \gamma_{kj} \tau_{ki} Y_k M, i = 1, \ldots, N_c, j = 1, \ldots, N_f \tag{12.11}$$

$$\sum_{i=1}^{N_p} Q_{ij} \leq X_j M, j = 1, \ldots, N_v \tag{12.12}$$

$$\sum_{i=1}^{N_p} \sum_{k=1}^{N_c} U_{ijk} \leq Z_j M, j = 1, \ldots, N_l \tag{12.13}$$

The constraint Eq. 12.7 ensures that sales do not exceed the demand. The sales-delivery balance is enforced by the constraint Eq. 12.8. The sales-supplies balance is enforced by Eq. 12.9 stating that products must be purchased from suppliers in order to sell them to the customers. Eq. 12.10 specifies that services should be selected to satisfy all the required information processing functions. Constraints Eq. 12.11–12.13 ensure that suppliers, providers and services, respectively, should be included in the supply chain if they perform any activities (e.g., products are supplied by the given supplier). Constraint (12.12) ties the physical and information flows by requiring that products cannot be physically delivered, if appropriate information services are not available.

12.4.3   Experimental

Experimental studies are conducted to demonstrate interdependencies between physical and information supply chain configuration decisions and to investigate impact of the weights w_1 and w_2 on the configuration results. In order to check the first aspect, the supply chain configuration is performed without taking into account the information flows (EXP1). Technically, it means that w_2 is set to zero and constraints (12.9) and (12.10) are ignored. The results of EXP1 are compared with an experiment (EXP2) where the physical and information flows are considered simultaneously. It is argued that the joint configuration has a significant impact on supply chain configuration, if different suppliers or 3PL providers are selected.

12.4.3.1 Design of Experiments

A test supply chain configuration problem is set up for experimental purposes. The dimensions of this supply chain are given by $N_p = 10$, $N_c = 30$, $N_v = 8$, $N_l = 3$, and $N_s = 10$. The services should provide seven functions. Services vary from highly specialized, providing just one function to aggregators providing all functions. Some of the services are available in all countries while others are limited just to selected countries. The demand is randomly generated. However, the average demand for certain products is country dependent, and some suppliers are able to produce these cheaper than others. The QoS characteristics are also randomly generated, though they are correlated with a number of functions the service provides (i.e., than more functions than worse performance). The model is solved using a commercially available mathematical programing software.

12.5 Results

Experiments EXP1 and EXP2 are carried out for the test supply chain. Figure 12.5 shows all supply chain unit evaluated during the configuration and the units selected are shaded. $P' = R - C_1 - C_2 - C_3$ measures the supply chain performance in each experiment. It can be observed that different supply chain configurations are obtained in both experiments. In EXP1, the supply chain is able to serve all

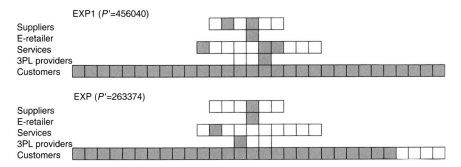

Fig. 12.5 The supply chain configurations obtained in experiments EXP1 and EXP2 (Grabis 2013)

customers. The electronic information flows are provided by a combination of three web services, and a 3PL provider which provides more uniform delivery costs around the world, is selected. In EXP2, the information processing is performed by a single aggregator, which covers all but four countries. One additional supplier is present in the results of EXP1 compared to the results of EXP2. This supplier is able to supply all products but it specializes in products most frequently ordered by customers in countries not served in EXP2. The supply chain performance is substantially affected by taking into account interdependencies between the physical and information flows for the given test supply chain.

The relative value of the weight factor w_2 characterizing the importance of QoS criteria in optimization is varied in order to evaluate sensitivity of results. The test supply chain used in the chapter is quite insensitive to this factor. The QoS criteria had significant impact on the configuration results only for values w_2 exceeding 10^5 (the cost related factors and quality related factors have vastly different scales).

12.6   Summary

Cloud computing, mobile devices, and sensing devices are transforming supply chain management. From the supply chain configuration perspective, the most obvious change is availability of new types of links among supply chain units as well as diminishing dependence on physical location of the supply chain units. Additionally, many supply chain units operate with entirely digitalized products and services. As a result, supply chain configuration assumes facets similar to the service orchestration and both physical and virtual supply chain units should be selected and connected together. A supply chain consisting of the physical and virtual supply chain units is referred to as cloud chain in this chapter.

A model for joint optimization of the physical and information flows in e-retailing supply chains has been elaborated. The model ensures that physical supply chain units have appropriate information processing capabilities at their disposal. The importance of the concurrent optimization increases along with a growing number of electronic services available over the Internet. The additive manufacturing or 3D printing can further process even more rapidly.

The formulated optimization model defines relationships between the physical and information flows and takes into account QoS requirements for efficient information processing. Preliminary experimental results show that the information flows indeed affect selection of appropriate physical supply chain units. However, the QoS requirements have minor impact of the supply chain configuration decisions for the test supply chain analyzed in the chapter. An alternative approach to including QoS criteria directly in the objective model would be the specification of minimum quality requirements in the form of constraints. That would also alleviate the problem of selecting appropriate weights for multi-criteria optimization. The QoS characteristics also have impact on customer demand which could also be represented in the optimization model.

References

Basole RC, Karla J (2011) On the evolution of mobile platform ecosystem structure and strategy. Bus Inform Syst Eng 3(5):313–322

Candido G, Barata J, Colombo AW, Jammes F (2009) SOA in reconfigurable supply chains: a research roadmap. Eng Appl Artif Intell 22:939–949

Grabis J (2013) Concurrent optimization of physical and information flows in supply chains. In: Proceedings of the 6th international workshop on information logistics, knowledge supply and ontologies in information systems, Poland, Warsaw, 23–25 Sept, 2013. pp 1–10

Hernandez B, Jimenez J, Martin MJ (2009) Key website factors in e-business strategy. Int J Inf Manag 29:362–371

Hong J, Suh E, Kim S (2009) Context-aware systems: a literature review and classification. Expert Syst Appl 36(4):8509–8522

Karakostas B, Katsoulakos T, Zorgios Y (2012) Towards an ICT platform for the European freight transport community. Int J Appl Logist 3(2):53–58

Klein R, Rai A (2009) Interfirm strategic information flows in logistics supply chain relationships. MIS Q 33:735–762

Metzger A, Franklin R, Engel Y (2012) Predictive monitoring of heterogeneous service-oriented business networks: the transport and logistics case. Annual SRII Global Conference, SRII, pp 313–322

Strunk A (2010) QoS-aware service composition: a survey. Proceedings of the 8th IEEE European Conference on Web Services, ECOWS 2010, pp 67–74

Swaminathan JM, Tayur SR (2003) Models for supply chains in e-business. Manag Sci 49:1387–1406

Trappey CV, Trappey AJC, Lin GYP, Liu CS, Lee WT (2007) Business and logistics hub integration to facilitate global supply chain linkage. Proc Inst Mech Eng B J Eng Manuf 221:1221–1233

Part IV
Applications

Chapter 13
Application in Hi-Tech Electronics Industry

13.1 Introduction

The hi-tech electronics industry produces a wide range of products. The best known examples are in consumer electronics, but around half of the produce goes to other types of end-products and B2B customers in diverse industries. While the consumer electronics sector is dominated by large OEMs, electronic parts are produced by a large number of smaller manufacturers. The specialist manufacturers form non-hierarchical collaborative supply chain networks (Scholz-Reiter et al. 2010) to produce integrated electronics end-products. These chains are characterized by short-product life cycles, high degree of customization and low margins. To respond to these pressures, a high degree of specialization can be observed in many hi-tech supply chains, where contract manufacturers offer their specialized knowledge and resources to product on-demand products.

There are manufacturers focusing on low value mass production parts and manufacturers doing their own R&D and providing high value specialized parts as well as those focusing on assembly (e.g., FoxConn[1]). In consumer electronics products, the final assembly cost often is just 5–10 % of the cost of the part used in the assembly.

The manufacturers are located around the world with the largest concentration in South East Asia, Europe, and the USA. The selection of parts manufacturers among other factors is driven by labor costs, scalability, proximity to other suppliers and customers, and quality of infrastructure. The clustering effect is particularly strong (Porter 1998). Recent experiences with part shortages due to natural disasters and other disruptive events have made many supply chains to rethink their reliance on the lean strategy and to make the supply chains shorter and more flexible. The main challenges affecting the electronics supply chains are improvement of supply chain collaboration (Siddiqui and Raza 2015) especially at the strategic and tactical

---

[1] http://www.foxconn.com/

© Springer Science+Business Media New York 2016

C. Chandra, J. Grabis, *Supply Chain Configuration*,
DOI 10.1007/978-1-4939-3557-4_13

levels, risk management, sustainability, demand planning, and digital supply chains. Demand planning implies that the contract manufacturers and B2B suppliers get more involved with the end-customers through various means of real-time monitoring. The aim is to provide services in a proactive manner and to sense the demand, rather than just observe it. The supply chain digitalization implies that products are augmented with different digital services as well as many production activities are becoming virtualized.

This chapter investigates characteristics of hi-tech electronics supply chains by considering the case of a company referred to as ET, which is a medium size contract manufacturer as well as a supply chain service provider. One of the main challenges faced by the company is associated with delivery of components used in manufacturing. In recent years, the estimated industry on-time delivery performance has deteriorated from around 95 % to 93 %. The delivery timeliness is affected by various disruptive events such as earthquakes, tornados and others (Chopra and Sodhi 2004). In order to evaluate these uncertainties, a simulation model for ET is constructed in this chapter. Simulation and analytical models have been successfully applied to study supply chain disruption in several investigations, e.g., Keramydas et al. (2015), MacKenzie et al. (2014), and Carvalho et al. (2012). MacKenzie et al. (2014) specifically focus on supply chain disruptions caused be the 2011 Japanese earthquake and tsunami and their model is used to evaluate risk management and post-disruption management strategies.

13.2 Case Description

ET is a contract manufacturer located in Latvia. It is a fast-growing group providing manufacturing services to business customers. The company runs two state-of-the-art technologically compatible plants—providing production capacity backup, supply reliability, and scalability of manufacturing processes. The service range covers the entire value chain from the design and industrialization phase to after-market services. The company's core markets are Baltic states, Finland, Sweden, Norway, the UK, and Denmark (Fig. 13.1).

While many supply chains are product centric, the mainstay of ET's supply chain is knowledge and technology and products are unique for every order. Therefore, the company also works with a large number of customers and suppliers. It continuously updates its customer portfolio to ensure that there are expected orders up to six month in advance. Similarly, it also selects suppliers dynamically according to the current requirements. The main supplier selection criteria are references and observations as well as test runs.

The products produced consist of three main types of components (Fig. 13.2):

1. Commodity components—readily available standard parts used in manufacturing of many end-products.

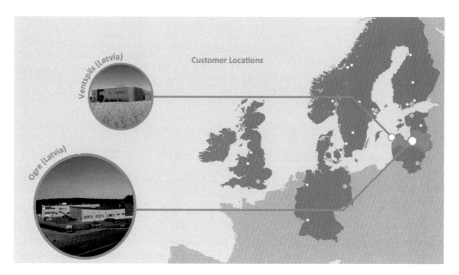

Fig. 13.1 ET geographical location

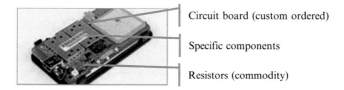

Fig. 13.2 A sample end product and main type of parts used in manufacturing

2. Custom ordered—must be ordered for every specific product, though alternative suppliers are readily available.
3. Specific components—components supplied just by limited number of suppliers (often a single supplier). These can be procured on order, or purchased from the catalog companies, though that usually costs more.

A majority of components are sourced on-demand. To ensure fulfillment of purchasing requisitions, the company has an advance agreement on prices and capacity reservation. Upon receiving a firm order, the company procures necessary materials. The materials have different delivery timeline which correlates with the type of materials, as illustrated in Fig. 13.3. In order to minimize inventory management costs, parts are sourced just-in time. The supply lead time for specialized parts is the longest one. The supply of commodity and custom parts is initiated taking into account the delivery slack available to receive the part on-time for end-manufacturing. The manufacturing is started once all parts are received. The process is completed by delivering the product to the customer. The company allows for a buffer when quoting the end-product delivery due date in order to account for supply and manufacturing uncertainties.

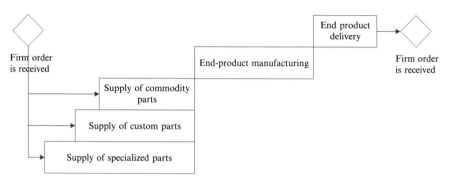

Fig. 13.3 The delivery timeline

The manufacturing order fulfillment time for approved design products is about 8 weeks (could increases to 28 weeks in the case of new products). Materials procurement time varies according to the type of material. The order quality and delivery time are agreed upon following the communication protocol given in Fig. 13.4. During the preliminary phase, the company forecasts its material requirements and informs suppliers about the expected orders. The suppliers send back their quotes specifying availability of products and their prices. At this point, both the company and the suppliers are yet to commit to firm orders. The sourcing phase starts once the manufacturing company receives firm orders from its customers. The manufacturing company commits itself to a certain end-product due date and quantity. Taking into account the material requirements, it places orders to suppliers. These orders specify the requested parts quantity and supply due data. The supplier sends back an order confirmation. It is possible that the due date promised by the supplier differs from the requested supplier's due date. The manufacturer decides upon accepting or rejecting the offer. In the case of accepting the offer, the supplier sends parts to the manufacturing company whenever these are ready for shipment. It is possible that the actual delivery date for parts is later than the promised date in the order confirmation because of unexpected disturbances. There is an important distinction between not offering the requested due date and not meeting the promised due date. In the former case, the manufacturer can take proactive measures to source the required parts within the allocated time. In the latter case, opportunities for proactive response are limited.

Differences between the required delivery date, confirmed delivery date and actual delivery date cause difficulties to meet the promised final product delivery date. Therefore, the company wants to evaluate its ability to meet the delivery date as well as to come up with strategies for dealing with the delays.

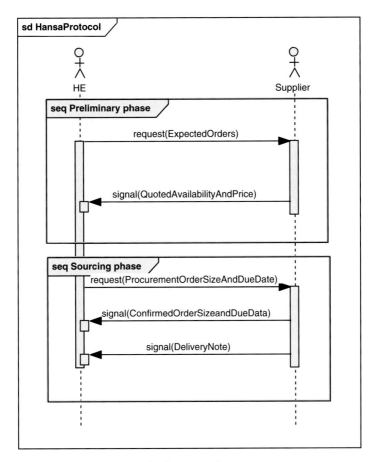

Fig. 13.4 Materials' ordering communication protocol

13.3   Scope Definition

The company pursues the efficiency strategy. It has mean manufacturing operations while relationships with customers and suppliers are agile. As part of the efficiency strategy, the company aims to ensure having as high utilization of its manufacturing facilities as possible. Paying attention to customer relationships and keeping the promised delivery performance are among the key competitive advantages of the company. Therefore, ability to evaluate feasibility of the promised delivery times are of high importance.

ET is a dominant unit of its supply chain and makes supply chain configuration decisions independently. The supply chain has a small number of fixed units since customers and suppliers are selected dynamically. Nevertheless, there is a portfolio of established customers and a pool of certified suppliers. There is a limited information sharing (the company gives suppliers in advance its own demand

Table 13.1 The supply chain configuration scope definition for the ET case

Scope parameter	Values
Objectives and criteria	Increase profit, increase capacity utilization, ensure reliable deliveries
Horizontal extent	Supply, manufacturing
Vertical extent	Strategic
Decisions	Sourcing policy
Parameters	Sourcing uncertainty, purchasing price
Processes and functions	Sourcing

predictions rather than information about the end-customer demand). The number of alternatives for selection of suppliers for commodity and customer parts is large. The summary supply chain scope definition is given in Table 13.1.

13.4   Conceptual Modeling

The conceptual modeling is performed to formally define the supply chain configuration problem in the case study. It is performed using the information modeling methods elaborated in Chap. 7. Figures 13.5 and 13.6 show the supply chain configuration objectives and the supply chain configuration concepts, respectively.

The goal view shows that the profit increase is the most important goal. In the supply chain configuration case considered, the goal is achieved by minimizing sourcing costs and increasing capacity utilization. Both goals are typical representatives of generic supply chain management objectives of cost optimization and improvement of asset management as identified in Chap. 7. The supplier selection objective facilitates the delivery reliability improvement because suppliers can be selected according to their on-time performance. The capacity utilization increase hinders delivery reliability at the manufacturing tier if too many manufacturing orders are booked at the same time.

The concept model defines main concepts relevant to the ET case. It explicitly shows that distinguishing among types of suppliers and types of materials is important. The concept model includes concepts for specifying contract suppliers and spot market suppliers as well as concepts for representing commodity, specialized and custom parts. The parts are traditionally shipped from suppliers to the manufacturer by air, which is represented by the Air Link object.

13.5   Simulation Model

Supply chain configuration evaluation experiments are designed on the basis of conceptual modeling. In order to evaluate impact of uncertainties on on-time delivery performance, simulation modeling is selected as the most appropriate

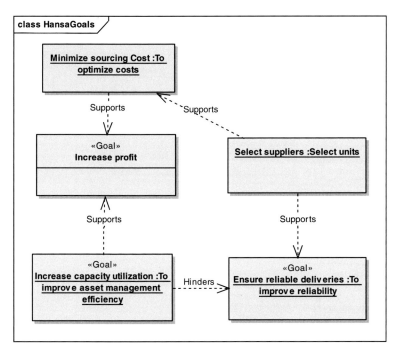

Fig. 13.5 The supply chain configuration objectives

method. The primary objective of the analytical evaluation is finding the probability to meet the promised delivery time as well as to evaluate an approach for dealing with uncertainty. This approach assumes that in the case of expected delays in parts' deliveries, they are procured at the spot market.

The main performance measures are the probability of meeting the promised delivery time, the expected delivery time and the sourcing costs. The probability p of meeting the promised delivery time $T_{promised}$ is expressed as:

$$p = P\left(\hat{T} \leq T_{promised}\right), \tag{13.1}$$

where $\hat{T}$ is the expected end-product delivery time. The expected end-product delivery time is expressed as:

$$\hat{T} = \max(T_{s1}, \ldots, T_{sn}) + T_m + T_d, \tag{13.2}$$

where T_{si}, $i = 1, \ldots, n$ is the actual supply time for the ith supplier, T_m is the end-product manufacturing time, and T_d is the end-product delivery time to the customer. The sourcing cost C is expressed as:

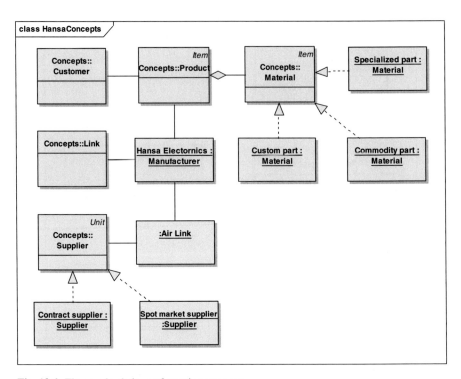

Fig. 13.6 The supply chain configuration concepts

$$C = c_1 \sum_{i=1}^{n} Q_i + c_2 \sum_{i=1}^{n} X_i + c_3 D \max\left(T - T_{\text{promised}}, 0\right), \qquad (13.3)$$

where c_1, c_2, and c_3 are the cost coefficients representing the regular purchasing price, spot market purchasing price, and late delivery penalty, respectively. Q_i is the quantity of materials sourced from the ith supplier at the regular price, X_i is the quantity of materials sourced from the ith supplier at the spot market, and D is the end-product demand.

The simulation model is built according to the simulation modeling principles presented in Chap. 9. Figure 13.7 shows the top level simulation model, where the first section represents the planning activities, the second section represents sourcing of commodity, custom and specialized components, respectively; and the third section represents the end-product processing. The sourcing activities are further elaborated in sub-model. Figure 13.8 shows a sub-model for sourcing of the specialized components. The sub-model is also developed using the principle of self-similarity, where sourcing operations are represented similarly for all suppliers. The top-level models shows that the quoted supply time T_{quoted} is provided by suppliers and the orders are sent out to suppliers. The manufacturing process can continue manufacturing operations of all components received. The specialized component sub-model shows that for every supplier T_{quoted} is compared with required supply time T_{required}. Meeting the required supply time should ensure

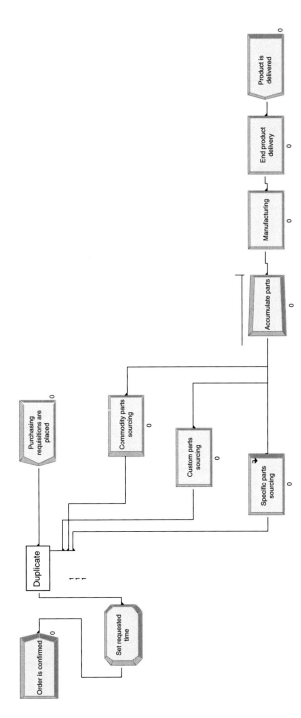

Fig. 13.7 The top-level simulation model

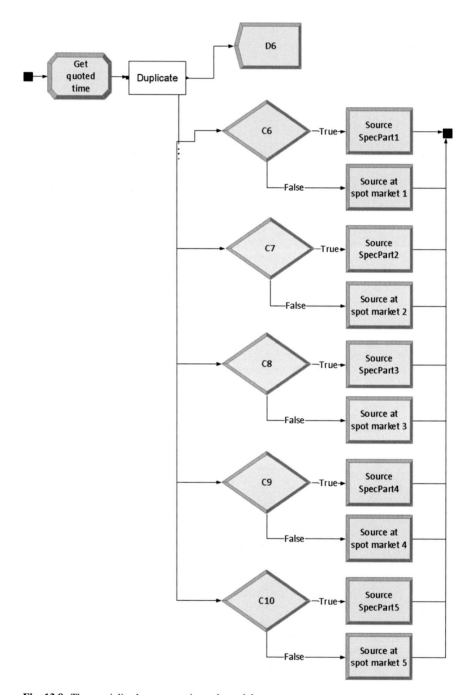

Fig. 13.8 The specialized parts sourcing sub-model

Table 13.2 Values of the experimental factors

Level	δ	β	α	$r = \alpha/c_3$
Low	1.1	0.1	1.2	1
High	1.3	0.5	1.5	5

on-time delivery of the end-product. If $\delta T_{quoted} \geq T_{required}$ then the order for the given component is placed at spot market instead of sourcing from the regular supplier. δ is called a switching threshold; given that the manufacturer allows a buffer to deal with delays and there is a tolerance level for late deliveries. The cost of parts in the spot market are higher $c_2 = \alpha c_1$, $\alpha > 1$, where α is the spot market purchasing threshold. The end-product demand is given by the customer. The execution time for all activities in the simulation model is log-normally distributed with the average value μ and standard deviation $\sigma = \beta\mu$, where β characterizes the level of activity execution uncertainty. In the case of sourcing activities, it characterizes the level of delivery uncertainty.

The simulation is performed at the strategic level and other factors influencing sourcing costs and delivery time are disregarded.

In order to evaluate the supply chain performance and to identify strategies for dealing with late deliveries, a set of experiments are conducted. A full factorial design of experiments is constructed for three experimental factors and one policy variable (Table 13.2). The experimental factors are: (1) a level of delivery uncertainty; (2) spot market premium; and (3) a ratio between the spot market premium and the late delivery penalty. The policy variable is the expected delivery lateness threshold δ at which the purchasing at the spot market is triggered.

The simulation model is developed in the ARENA modeling environment. The simulation is performed for 100 replications.

13.6 Experimental Results

The expected delivery performance is initially evaluated. The delivery time in the model depends only upon the level of delivery and the switching threshold while the spot market premium and late delivery penalty affect only the sourcing cost. Figure 13.9 shows the distribution of the expected delivery time as estimated over 100 simulations. The promised delivery time $T_{promised} = 49$ days. It can be observed that increasing β significantly affects the delivery time, while lower switching threshold helps improve on-time delivery performance. In the case of low delivery uncertainty, for the given end-product delivery buffer, the on-time delivery probability is 1. In the case of high delivery uncertainty, the on-time delivery probability is 0.97 and 0.89 for low and high values of the switching threshold, respectively. Therefore, the high supply lateness tolerance does not allow to achieve the average industry wide on-time delivery performance.

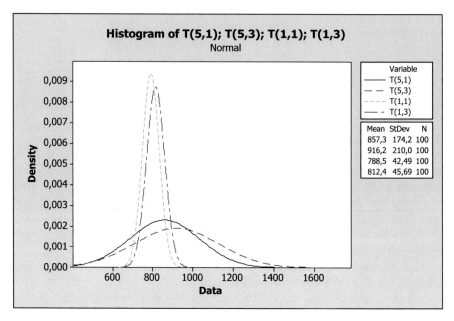

Fig. 13.9 The distribution of the expected delivery time

The impact of the switching threshold on the sourcing cost and interactions between the experimental factors are shown in Fig. 13.10. The sourcing cost is most significantly affected by the level of delivery uncertainty and spot market premium. The high level of delivery uncertainty increases the number of cases when parts are ordered in the spot market and at the same time increases the late delivery penalty. In the case of low level of delivery uncertainty, having the higher lateness tolerance is advantageous compared with too early switching to the spot market (Fig. 13.10a). There are no significant interactions among the spot market premium and the switching policy (Fig. 13.10b). The switching threshold has an opposite effect on the cost depending upon the r (Fig. 13.10c). The low lateness tolerance causes heavy purchasing on the spot market to avoid late delivery but for low r, spot market purchasing premium outweighs reduction in late delivery penalty. The opposite effect is observed in the case of high lateness tolerance.

13.7 Summary

This chapter describes the case study of supply chain configuration at an electronics manufacturing company. The main attention was devoted to suppliers' relationships management, in order to adopt appropriate policies for supplier selection. The main supplier selection driver was the impact of supply reliability on promised

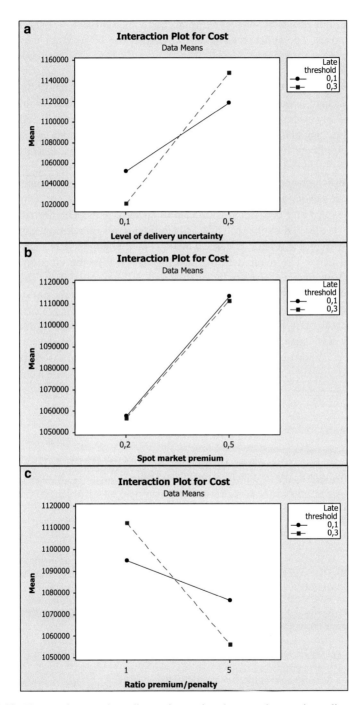

Fig. 13.10 The sourcing cost depending on interactions between the sourcing policy and experimental factors: (**a**) level of delivery uncertainty; (**b**) spot market premium; and (**c**) ratio between premium and penalty

end-product delivery time. The simulation modeling approach was used for evaluation purposes. The conceptual model developed can be used for exploring the supplier selection problem as well as for investigation of other supply chain configuration issues at the company.

The case study revealed that electronics supply chains are highly flexible, and supply chains are frequently established on project-to-project basis for manufacturing, particularly custom-built end-products. In every project, the main configuration variables are associated with supplier selection while logistics operations are streamlined without having to use multitiered storage facilities. Despite the high level geographical distribution in the electronics supply chains the transportation is relatively efficient because the parts and end-products are high value, low mass products. However, a major issue is that the supply chains are vulnerable to disruptive events and occasional shortages of specific parts. The decision-making complexity is affected by the type of parts required in manufacturing and each type requires a different sourcing strategy. The most challenging task is sourcing of the specialized parts. The manufacturing margins are low in the electronics supply chains, and companies compete by quality and delivery reliability. The dynamic modeling methods, such as simulation are the most useful methods for investigating the electronics supply chains.

The characteristic feature of contract manufacturing companies is that by handling their own manufacturing operations they acquire significant supply chain management expertise and are able to offer this expertise as a service to other companies.

References

Carvalho H, Barroso AP, MacHado VH, Azevedo S, Cruz-Machado V (2012) Supply chain redesign for resilience using simulation. Computers and Industrial Engineering 62(1):329–341

Chopra S, Sodhi MS (2004) Managing risk to avoid: supply-chain breakdown. MIT Sloan Manage Rev 46(1):53–61

Keramydas C, Tsiolias D, Vlachos D, Iakovou E (2015) A simulation methodology for evaluating emergency sourcing strategies of a discrete part manufacturer. Int J Data Anal Tech Strat 7 (2):141–155

MacKenzie CA, Barker K, Santos JR (2014) Modeling a severe supply chain disruption and post-disaster decision making with application to the Japanese earthquake and Tsunami IIE transactions. Inst Ind Eng 46(12):1243–1260

Porter ME (1998) Clusters and the new economics of competition. Harvard Bus Rev 76(6):77–90

Scholz-Reiter B, Heger J, Meinecke C, Rippel D, Zolghadri M, Rasoulifar R (2010) Supporting non-hierarchical supply chain networks in the electronics industry. In: Pawar K, Canetta L, Thoben K, Boer C (eds) Proceeding of the 16th international conference on concurrent enterprising. Centre for Concurrent Enterprise, Nottingham University Business School, Nottingham, pp 1–6

Siddiqui AW, Raza SA (2015) Electronic supply chains: status & perspective. Comput Ind Eng 88:536–556

Chapter 14
Application in ICT Distribution

14.1 Introduction

The Information and Communication Technology (ICT) industry is a diverse industry of major economic importance. Computer hardware is an important part of this industry. Sales of the computer hardware were around 700 billion in 2014,[1] and the market continues to grow rapidly. It is strongly affected by the trend of computing consumerization which puts pressure on supply chain responsiveness. The ICT supply chains must respond to rapidly changing technological trends, characterized by a high level of competition and global distribution. Therefore, supply chain agility and reconfigurability are of high importance in this domain.

Multiple players are involved in ICT supply chains, including manufacturers, distributors, retailers, and end-customers. The manufacturers most prominently feature contract manufacturers and original equipment manufacturers. The ICT supply chain from the perspective of the contract manufacturer is described in Chap. 13. There are different types of retailers and end-customers, and the omnichannel approach is widespread (Piotrowicz and Cuthbertson 2014).

The nature of ICT products enables a significant share of direct sales from manufacturers to end-customers. However, ICT distributors also play a major role. They provide services to both customers and manufacturers by providing an extensive product portfolio and a broad line of logistics services (Balocco et al. 2012). The typical core services provided to the customers are high availability, bulk breaking, financing, technical expertise and pre-sales product information, order consolidation, and delivery. The main additional or optional services provided to the manufacturers and other vendors of the ICT products are market knowledge, demand generation, local logistics, and after-sales services (GTDC 2013).

---

[1] http://www.gartner.com/technology/research/it-spending-forecast/

© Springer Science+Business Media New York 2016
C. Chandra, J. Grabis, *Supply Chain Configuration*,
DOI 10.1007/978-1-4939-3557-4_14

In order to provide insights into operations of ICT supply chains from the distribution perspective, this chapter presents a case study of a major Eastern European ICT distributor and its supply chain. The case is analyzed following the integrated supply chain configuration methodology and the main focus areas of this investigation are supply chain visualization along with spatial analytics, as well as dynamics of new product introductions. The case is elaborated on the basis of interviews with company representatives in 2013 though some of the factual information used is not company specific.

14.2   Case Description

A major East European ICT distributor is headquartered in Riga, Latvia. Its annual sales exceed one billion USD. It serves ten Eastern European countries and has actively expanded its business in Central Asia. The main two courses of action are expertise in a variety of solutions and services and wholesale of computer and electronic products. The company operates its distribution network and procures products around the world; from South East Asia in particular. It has cooperation and distribution agreements with more than 70 leading international manufacturers. The product portfolio includes more than 20,000 product titles. The company has more than 7000 customers ranging from retailers to local computer manufacturers and system integrators. For handling day-to-day operations the company uses a customers and suppliers relationships management system. The system provides access to all services provided by the company. Its main functions are purchasing and sales processes, marketing processes, transit tracking, financing; as well as information services on new products and promotional and technical events.

Figure 14.1 shows a pictorial representation of company's distribution network. Five logistics hubs are located in Amsterdam, Rotterdam, Helsinki, Dubai, and Riga. Local warehouses are located in each of the markets of operation. The company operates its warehouses by itself as well as uses logistics service providers. Delivery of goods is outsourced to international freight forwarding companies such as DHL, DSV, TNT, Schenker, and others. The standard types of delivery contracts are used depending upon suppliers' requirements and customers' preferences.

The company's strategy is to provide end-to-end services to its customers and to offer the whole range of products. At the same time, the company focuses on offering high-end components and devices, and introducing latest technologies competing primarily on quality and services rather than price.

The company has six major types of product: Desktop Solutions, Mobile Solutions, Consumer and Multimedia, Server, Storage & Security Solutions, Software, and Smartphones and Tablets. Products are segmented as trend products, mainstream products, and commodity. The product segmentation is one of the key drivers for supply chain management. The company's distribution strategy also depends upon product distribution strategies pursued by suppliers.

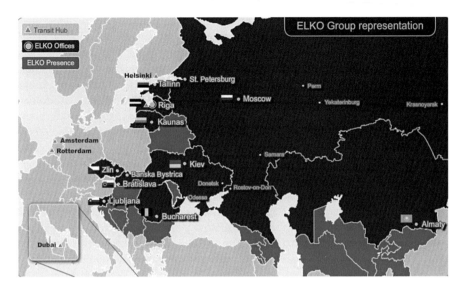

Fig. 14.1 The company's supply chain

The supply chain configuration is relatively stable and it is well aligned with the global electronic supply chains. Atzema (2001) and APEC (2013) shows that the global electronics supply chains cluster around certain locations where facilities of major manufacturers, wholesalers, and 3PL providers are located. For instance, Amsterdam is one of such hubs where majority of large companies have their logistics centers.

The main points of supply chain configuration variation are supplier selection for commodity products and expansion in new markets, mainly, by introducing new products. These points of variation are main concerns of supply chain configuration decision-making. The company currently relies mainly on managerial decision making methods while increasing supply chain complexity calls for a more formal approach, and the integrated supply chain configuration methodology is applied in this test case.

The following sections describe the main activities performed in the framework of the integrated supply chain configuration methodology. The emphasis is devoted to analysis scope definition, conceptual modeling, and analytical modeling.

14.3 Scope Definition

The company pursues the growth strategy. It fosters long-term collaboration with its partners and at the same time prefers the service-oriented supply chain strategy with regard to structuring and operating its distribution network implying that the company relies on the third party service providers for many of the logistics services.

To exemplify application of the supply chain configuration methodology, it is assumed that the company uses the methodology to investigate supply chain configuration options. As a part of its growth strategy, the company evaluates offering new products in one of its markets. It uses the integrated configuration methodology to better understand the impact of this decision on the supply chain integrity and to evaluate its competitive position for these new products. Additionally, the company would like to have means for quantitative analysis of different aspects of new product introduction from the supply chain configuration perspective.

The decision making circumstances are characterized by the dominant unit, which is the company itself. Obviously, it is restricted by a number of factors including partners' requirements (some suppliers do not allow distribution of their products in certain markets) but the configuration decisions are made exclusively internally. The initial state is the existing supply chain with fixed supply chain units. There is limited information sharing and some historical records are available. The number of alternatives to be evaluated is small. The company already operates in the target market but it intends to offer new products to existing customers. These products are provided by a single supplier. The company has to choose the distribution route, i.e., whether products are delivered via existing distribution facilities or direct supplies can be used. It also negotiates with the supplier delivery conditions. The standard INCOTERM (Malfliet 2011) types of contract are available. These contracts define allocations of costs and risks associated with delivery of products between the supplier and the customer. Different configurations and contracts can be used for different products and customers. The purchasing and handling costs depend upon both the configuration and the type of contract used.

The supply chain configuration scope definition is given in Table 14.1. The overall goal of the configuration is to evaluate offering products in a new market. The main evaluation objective is to minimize distribution cost. The choice of the distribution configuration and the contract type are the key decision variables. Decisions are strongly affected by supplier setup cost and fixed facility usage cost.

Table 14.1 The scope definition

Scope parameter	Values
Objectives and criteria	Offer products in a new market, minimize distribution cost
Horizontal extent	Distribution, supply
Vertical extent	Strategic
Decisions	Establishing the delivery route Selection of the delivery contract type
Parameters	Setup cost, fixed facility, purchasing cost, handling cost
Processes and functions	Procurement

14.4 Conceptual Modeling

In order to formally define the supply chain configuration problem, the goal, concepts and process models are developed for the case following the guidelines provided in Chap. 7. Figure 14.2 shows the case level goal model, which includes only goals that are most relevant to the company's supply chain configuration. The model shows To minimized distribution cost and To configure ICT distribution supply chain as two central goals for the case study. The new market is entered for products to increase flexibility, though that is affected by supplier's strategy on offering their products in the given market. Entering into a new market is associated with initial setup costs due to establishing supporting organizational structures, working procedures and communication channels. The company wide strategy prescribes that existing distribution facilities should be used, if possible. The supply chain configuration decisions to be made are concerning selection of delivery route and contract type.

The concept model at the case level (Fig. 14.3) contains two main types of elements: (1) relevant domain level concepts; and (2) case specific instances of the domain level concepts. The domain level concepts included in the concept model are Supplier, Product, Customer, Distribution Center, and Link while some concepts, for instance, Manufacturer are not included because they are not relevant to the particular case (i.e., manufacturing is not considered in the given supply chain). Associations among these concepts are used to define general relationships among different constituent parts of the supply chain. These associations are subsequently important to define data structure and analysis models. The Link concept is used to represent a connection between supply chain units. For

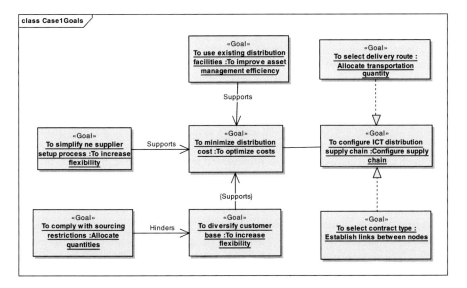

Fig. 14.2 The case level goals

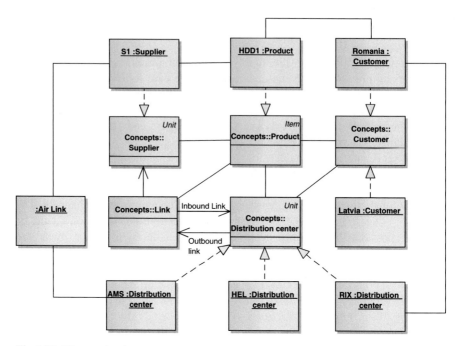

Fig. 14.3 The case level concepts

instance, it shows that products are delivered from suppliers to distribution centers via a given link. There are two associations between Link and Distribution Center to show that a link binds together two distribution centers.

The case specific instances of the domain level represent individual supply chain elements such as particular products or suppliers. The supply chain can include a large number of individual elements and only the most important elements are represented separately. In this case, instances are explicitly modeled for supplier S1, product HDD1, customers in Latvia and Romania and all five distribution centers the company operates. The associations among the instances show that the HDD1 product is supplied by the S1 supplier, the HDD1 product is intended for customers in Romania and that the product is delivered to the customer from the RIX distribution center via the AMS distribution center. The link between the AMS and the RIX distribution centers is represented using the generic link concept. The link between S1 and AMS is represented using the Air Link object because deliveries are made by air only.

Figure 14.4 shows the case level process model. The process model includes both generic supply chain concepts and specific supply chain objects. The generic concepts are used to show that products are shipped from suppliers to distribution centers. The process model shows that in general products can be shipped to any of the distribution centers as well as customers can be served from any of the distribution centers. However, at the level of specific supply chain objects, there

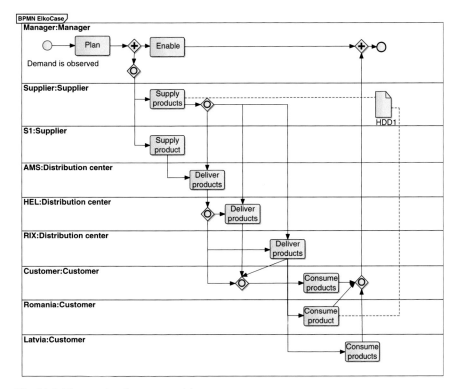

Fig. 14.4 The case level process model

are particular shipment and delivery restrictions. The model shows that supplier S1 ships only to AMS, and customers in Romania are served from the distribution center in Latvia. Therefore, products from AMS are first delivered to RIX and only then to the final customers. These kind of restrictions occur in practice because suppliers often have their logistics hubs at certain locations and they route all activities through these hubs.

The experimental planning is performed on the basis of information accumulated during the conceptual modeling. It is decided that a spatial analysis should be conducted using supply chain visualization and data fusion as well as an optimization model should be developed to select the delivery route and the contract type.

Figure 14.5 offers graphical representation of the supply chain network. It shows that supply chain units are fixed and the configuration objective is to establish appropriate links among the units. The dashed line is used to represent the links in question. All suppliers involved in the supply chain configuration initiative are represented as a group and there are various options for connecting the customers with other supply chain units.

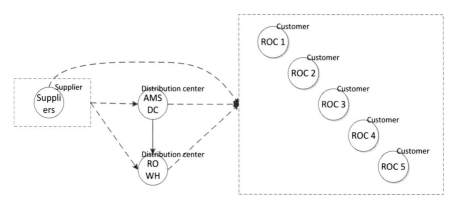

Fig. 14.5 Graphical representation of the supply chain network

14.5   Optimization Model

The configuration problem is to select the distribution route and the contract type in order to minimize distribution costs. The distribution route is defined as a chain from the supplier to the customer, and four distribution routes are evaluated:

- Complete route (CR)—the delivery chain is a sequence of supplier, AMS distribution center, ROM warehouse, customer.
- Hub base route (HBR)—the delivery chain is a sequence of supplier, AMS distribution center, customer.
- Local route (LR)—the delivery chain is a sequence of supplier, ROM warehouse, customer.
- Direct route (DR)—the delivery chain is a sequence of supplier, customer.

These routes are graphically illustrated in Fig. 14.6.

The complete route takes advantage of the existing distribution facilities while the direct route attempts to straighten the delivery process. However, the direct route is not available for all products and all customers. The direct route also cannot take advantage of bulk processing of shipments, resulting in higher handling costs. The existing distribution facilities are shared with all other products that the company distributes. Therefore, the fixed facility usage cost is charged as a fraction of the total facility costs depending upon the usage intensity.

There are four types of INCOTERMS contracts considered (Fig. 14.7):

- EXW—the seller makes the goods available at his/her premises. The buyer is responsible for uploading and all other delivery activities.
- CIF—the seller delivers goods to the port of import, where the buyer takes over further responsibilities.
- DAP—the seller is responsible for delivery at the destination though the buyer should clear the goods for import.
- DDP—the seller is responsible for all delivery and custom clearance activities and the buyer receives the goods at his/her premises.

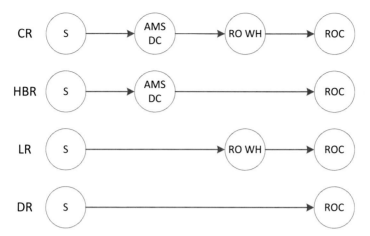

Fig. 14.6 Alternative distribution routes

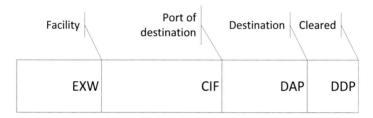

Fig. 14.7 Types of the contract considered

The purchasing price and the handling costs for the company depend upon the contract type. Generally, the purchasing price is lower for contracts with limited supplier's responsibility and the handling costs are higher for these contracts.

14.5.1 Model Description

A mathematical programming model is formulated to solve the configuration problem. The notations used in the model are defined in Table 14.2.

14.5.1.1 Objective Function

The supply chain configuration is established by selecting the most appropriate routes and contract conditions for delivering products. It is optimized to minimized total costs (Eq. 14.1). The total costs consist of purchasing costs (Eq. 14.2), handling costs (Eq. 14.3), supplier setup cost (Eq. 14.4), and fixed cost (Eq. 14.5):

Table 14.2 Notations

Notation	Definition
Indices	
i	Products, $i = 1, \ldots, I$
j	Customers, $j = 1, \ldots, J$
k	Contracts, $k = 1, \ldots, K$
l	Configurations, $l = 1, \ldots, L$
Parameters	
d_{ij}	Demand
g_{ik}	Purchasing price of the product for the given type of contract
h_{ik}	Handling price of the product for the given type of contract
γ_{kl}	Binary indicator of compatibility between the type contract and configuration (equals 1 if the contract type can be used together with the given configuration and 0 otherwise)
δ_k	Contract type setup cost
ω_l	Processing cost coefficient for the given configuration
λ_l	Share of facilities fixed cost attributed to the given configuration
p_i	Product value
f	Facilities fixed cost
t	Minimum turnover level to allow direct deliveries to the customer
T	Total estimated turnover
M	A large constant
Decision variables	
C_k	Binary indicator of type of contracts used
X_{jk}	Binary indicator of type of contracts used by the customer
U_{jl}	Binary indicator of type of configurations used for serving the customer
Z_{kl}	Binary indicator of type of configuration combined with the contract
Y_{ijkl}	Binary indicator of using given type of contracts and configurations for delivering the product to the customer
S_{ijkl}	Quantity of products delivered to the customer by using given type of contracts and configurations

$$\min TC = c_1 + c_2 + c_3 + c_4 \tag{14.1}$$

$$c_1 = \sum_{i=1}^{I} \sum_{j=1}^{J} \sum_{k=1}^{K} \sum_{l=1}^{L} g_{ik} S_{ijkl} \tag{14.2}$$

$$c_2 = \sum_{i=1}^{I} \sum_{j=1}^{J} \sum_{k=1}^{K} \sum_{l=1}^{L} \left(h_{ik} + \omega_l p_i \right) S_{ijkl} \tag{14.3}$$

$$c_3 = \sum_{k=1}^{K} \delta_k C_k \tag{14.4}$$

$$c_4 = \frac{f}{T} \sum_{i=1}^{I} \sum_{j=1}^{J} \sum_{k=1}^{K} \sum_{l=1}^{L} \lambda_l g_{il} S_{ijkl} \tag{14.5}$$

The handling cost calculation takes into account processing coefficient, which is configuration specific and adjusts the base handling cost (i.e., handling is more expensive if the buyer assumes more responsibilities). The handling cost includes processing at storage locations as well as transportation. The supplier setup cost concerns costs associated with setting up delivery procedures (e.g., IT support), which depend upon the type of contract used. The setup cost is higher if the supplier is responsible for more delivery activities.

The fixed cost calculation takes into account that the configuration has different facilities' usage intensity; and the coefficient λ_l is smaller, if fewer facilities are used. However, this might be counterintuitive to the need for more efficient usage of shared facilities. This issue will be explored in experimental studies.

14.5.1.2   Constraints

The objective function is minimized subject to following constraints. Equation (14.6) imposes that a contract type should be set up if it is to be used by at least one customer. Similarly, the contract type should be supported, if a customer uses it for any of the products and the routes (Eq. 14.7). Equation (14.8) balances supply and demand. Equations (14.9)–(14.11) set a type of configuration and contract used for the planned deliveries.

$$\sum_{j=1}^{J} X_{jk} < MC_k, \forall k \tag{14.6}$$

$$\sum_{i=1}^{I} \sum_{l=1}^{L} S_{ijkl} < MX_{jk}, \forall j, k \tag{14.7}$$

$$\sum_{k=1}^{K} \sum_{l=1}^{L} S_{ijkl} = D_{ij}, \forall i, j \tag{14.8}$$

$$S_{ijkl} < MY_{ijkl}, \forall i, j, k, l \tag{14.9}$$

$$\sum_{i=1}^{I} \sum_{k=1}^{K} S_{ijkl} < MU_{jl}, \forall j, l \tag{14.10}$$

$$\sum_{i=1}^{I} \sum_{j=1}^{J} Y_{ijkl} < MZ_{kl}, \forall k, l \tag{14.11}$$

$$g_{ik} S_{ijkl} > t Y_{ijkl}, l = 4, \forall i, j, k \tag{14.12}$$

$$\gamma_{kl} < Z_{kl}, \forall k, l \tag{14.13}$$

Equation (14.12) imposes that direct deliveries are allowed only if the minimum shipment value requirements are satisfied. Allowed combinations of configurations and contracts are constrained by Eq. (14.13), e.g., direct shipment is not compatible with contract EXW.

14.5.2   Experimental Evaluation

The optimization model is solved to select the appropriate configuration and contract type. Additionally, the impact of fixed costs allocation on configuration results is investigated.

14.5.2.1   Experimental Data

The number of products I is five and the number of customer J is five as well. Products range from high to low value and their characteristics are listed in Table 14.3. The table lists base purchasing price, which is adjusted depending on the type of contract used. It correlates with the level of responsibilities the supplier has. Similarly, the base handling cost is transformed into contract dependent handling price by multiplying with the contract specific handling multiplier (Table 14.4). The contract specific price is negatively correlated with the level of responsibilities the supplier has.

Four aforementioned delivery configurations and contract types are considered. The contract types (Table 14.4) differ by their setup costs and handling multiplier used to adjust the average handling price for products listed above. The configurations (Table 14.5) differ by their handling cost multiplier and fixed cost allocation multiplier. These values are estimated by an expert.

Table 14.3 Characteristics of the products

Product	Demand	Base price	Base handling cost
P1	44	40	10
P2	226	8	2
P3	180	1.5	0.375
P4	2	0.5	0.125
P5	68	0.3	0.075

Table 14.4 Characteristics of the contract types

Contract type	Supplier setup cost	Handling multiplier
EXW	0	1
CIF	10	0.65
DAP	30	0.25
DDP	40	0.1

Table 14.5 Characteristics of the configurations

Configuration	Handling cost multiplier	Fixed cost multiplier
CR	1	1.5
HBR	1.2	1
LR	1.25	0.9
DR	1.5	0

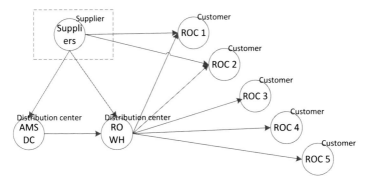

Fig. 14.8 The optimized supply chain network

Table 14.6 Count of delivery modes used for different products and customers		Configuration			
	Contract	CR	HBR	LR	DR
	EXW	3	0	0	0
	CIF	0	0	0	0
	DAP	0	0	3	2
	DDP	0	0	16	1

14.5.2.2   Results

The optimization results are obtained for five customers procuring five different types of products and different configurations and a different combination of delivery contract and configuration can be selected for every product and customer, resulting in 25 potential delivery modes. Figure 14.8 shows a graphical representation of the supply chain network obtained as a result of optimization. A link in the figure is present, if it is used by at least one product. Only the second configuration relying on direct deliveries from DC to customers is not used (see also Table 14.6). The most frequently used configuration is LR (i.e., suppliers delivers directly to the regional warehouse, which distributes products to customers).

LR configuration usually is used together with the DDP contract (Fig. 14.9). The supplier delivers products to the regional warehouse and clears duties and taxes under the DDP contract. The focal company is responsible for distributing products to customer locations. One interpretation of the results is that if suppliers are willing and able to perform all paperwork associated with deliveries, then the importance of routing deliveries via the central DC diminishes. CR usually is used if minimum delivery requirements cannot be met (see also Fig. 14.11). In this case, the EXW type of contract is used. The focal company is responsible for all logistical operations and products are routed through the main distribution center and the regional warehouse. The DAP contract is used together with the DR configuration. The supplier sends the products directly to the customer and the focal company processes duties and taxes on behalf of the customer.

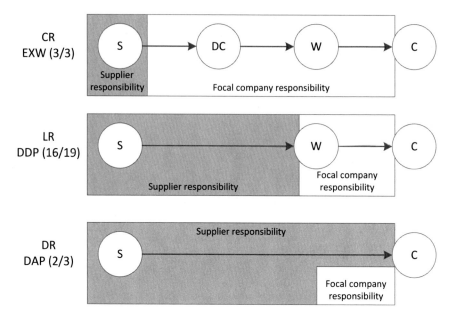

Fig. 14.9 Relationships among contracts and configurations. The *numbers* in parenthesis indicate usage of the specified type of contract out of all occasions of using the configuration

Table 14.7 Count of configurations used according to the product

Products	CR	HBR	LR	DR
P1	0	0	4	1
P2	0	0	3	2
P3	0	0	5	0
P4	3	0	2	0
P5	0	0	5	0

Table 14.7 shows that the LR configuration is used for all product groups. The exceptions occur in the case when customer require small amounts of the product as evident for the P4 product, which is low value, low volume product or in the case of P1 and P2, which are high value or high volume products.

The contract type and configuration selection is mainly driven by a trade-off between procurement cost and processing cost. To illustrate this trade-off, optimization is performed by restricting the choice of contract to just one option, i.e., the specified contract should be used regardless of customer, product and configuration. Optimization results depending on the type of contract imposed are shown in Fig. 14.10. It can be observed that the purchasing cost increases as the supplier assumes more responsibilities and includes additional expenses in the sales prices. At the same time, the handling cost decreases for the distribution company. Given that the distribution operations are one of the key competences of the distributor, it

Fig. 14.10 Handling and procurement costs according to the type of contract

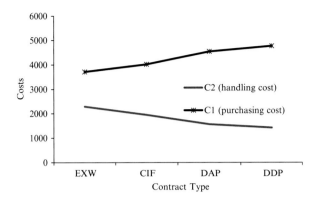

Table 14.8 Fixed costs evaluation scenarios

Scenario	CR	HBR	LR	DR
EXP0	1,5	1	0.9	0
EXP2	0.85	0.85	0.85	0.85
EXP3	2	0.7	0.7	0
EXP4	1.2	1.05	0.9	0.25

would prefer contracts with lower purchasing prices. However, in this case the DDP contract is favored. That could be explained by insufficient experience for supporting the new market or too high overhead of using the established facilities. The latter issue is investigated in the following subsection.

14.5.2.3 Impact of Fixed Costs

Configurations CR, GBR, and LR utilize existing logistical infrastructure to deliver products to the customer. There are certain fixed costs associated with these facilities. The model assumes that fixed costs are incurred as a fraction of the total cost depending on processing volume, i.e., if more products are shipped via the main distribution center then a larger share of the fixed costs should be paid by Romanian operations. Obviously, incentives of using the centralized facilities depend upon the usage conditions, which are formally expressed by λ_l share of facilities fixed cost attributed to the given configuration.

In addition to the base scenario, three scenarios for fixed cost allocation are defined (Table 14.8). The base scenario (EXP0) describes the most plausible cost allocation scenario. EXP2 assumes that fixed costs are allocated without respect to the configuration used. EXP4 has relatively even distribution of costs among the configurations and encourages use of existing facilities. EXP3 has the largest allocation variability and might be deterrent towards using CR.

The optimization results for different scenarios clearly indicate the impact of fixed cost allocation on configuration decisions (Table 14.9). CR is the most widely

Table 14.9 Optimization results and count of configurations used

Scenario	TC	CR	HBR	LR	DR
EXP0	7516	3	0	19	3
EXP2	8262	25	0	0	0
EXP3	7148	0	19	3	3
EXP4	7895	3	0	19	3

Fig. 14.11 The cost breakdown according to the fixed cost allocation scenario

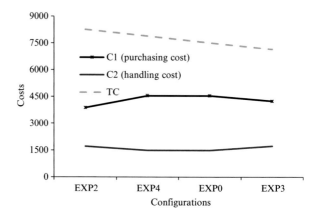

used configuration for EXP2 because cost allocation favors usage of shared distribution facilities reducing the handling cost. However, the total cost is the highest for EXP2, as there is a limited flexibility to find the right product to configuration mix (e.g., larger handling costs for DR are not contra-balanced by reduction of fixed costs). The lowest cost is achieved in EXP3 with the widest allocation of the fixed costs. The LR configuration is favored in the case of slightly varied allocation of the fixed costs as in the case of EXP0 and EXP4.

The cost breakdown for different fixed cost allocation scenarios is graphically illustrated in Fig. 14.11. The lowest cost corresponds to an optimal trade-off between the handling and purchasing costs. The CIF contract type is the most widely used contract type for EXP3 and using the centralized DC facilities (HBR) allows keeping handling costs relatively low while capitalizing on the low purchasing costs offered by the supplier for CIF type contracts.

14.6   Summary

The supply chain configuration methodology has been applied to analyze configuration challenges at the ICT distribution company. The company's supply chain has evolved in a step-by-step manner and the network development decisions have been made on a case-by-case basis. Consequently, the supply chain configuration models developed in the case study help to establish a systematic basis for further supply chain evolution and provide means for communicating the supply chain configuration decisions. The descriptive models are of particular value for these purposes.

Selection of the most appropriate delivery route for customers in a new market was an immediate concern of the configuration initiative. That also led to exploration of relationships between utilization of the common infrastructure and the allocation of fixed costs. The main distribution centers are the backbone of logistical operations while efficiency of local units facilitates further expansion. The quantitative analysis suggests that there should be flexibility in the allocation of fixed costs and achieving a trade-off between handling costs and fixed costs, leading to minimization of total costs. The required flexibility also implies that cost sharing among different product lines and customer accounts might be necessary to minimize the overall supply chain cost.

The model for evaluation of different delivery modes as combination of delivery contracts and delivery routes is a novel type of supply chain configuration model potentially valuable for other companies providing logistics services.

References

APEC (2013) Global supply chain operation in the APEC region: case study of the electrical and electronics industry. http://publications.apec.org/publication-detail.php?pub_id=1431

Atzema OALC (2001) Location and local networks of ICT firms in the Netherlands. J Econ Soc Geogr 92(3):369–378

Balocco R, Ghezzi A, Rangoneand A, Toletti G (2012) A strategic analysis of the European companies in the ICT sales channel. Int J Eng Bus Manag 4(6):1–10

GTDC (2013) Understanding the technology distribution business. http://www.gtdc.org/research/Understanding-the-Technology-Distribution-Business.pdf

Malfliet J (2011) Incoterms 2010 and the mode of transport: how to choose the right term. Proceedings of management challenges in the 21st century: transport and logistics: opportunity for Slovakia in the era of knowledge economy, pp 163–179

Piotrowicz W, Cuthbertson R (2014) Introduction to the special issue information technology in retail: toward omnichannel retailing. Int J Electron Comm 18(4):5–16

Chapter 15
Application in Health Care

15.1 Introduction

The US health care industry accounted for 17.1 % of the US economic output in 2013 according to the World Health Organization,[1] whereas EU averages 8 %. Various studies of this industry point to lack or failure of basic quality-control procedures, and misalignment among consumer needs, payers, and provider services, as primary causes for building waste into industry management practices (Kaplan 2012).

Pressures on the industry have fostered innovation in the design of services and organizations. Most of the innovations have targeted cost reductions in key functions, including logistics. The industry must find a flexible delivery enterprise that has substantial capital and is capable of efficient operations. This means effective management of a broad range of processes with diverse measures, from medical outcomes to cost of tissue paper. The health care sector of US economy faces several challenges, such as cost containment, outdated information management systems, and mergers and/or acquisitions. The need to cut costs and compete has led to mergers and acquisitions in health care industry. Such consolidations have created new organizations made up of very different entities which are not as integrated as they should be. Due to competition, it has become imperative that enterprises seamlessly and efficiently provide and manage services (including purchase and delivery of supplies to the final user) across entities and continuum of care, both now and in the future. McKone-Sweet et al. (2005) review main challenges in health care supply chains including short product life cycles, unpredictable patient flow, lack of standardization, outdated information technologies, and inadequate knowledge on supply chain management issues.

The principal participants in the US health care supply chain include: manufacturers (drugs, medical equipment, and hospital medical supplies), distributors,

---

[1] http://www.who.int/countries/usa/en/.

© Springer Science+Business Media New York 2016
C. Chandra, J. Grabis, *Supply Chain Configuration*,
DOI 10.1007/978-1-4939-3557-4_15

medical service providers, medical groups, insurance companies, government agencies (such as the Health and Human Services department), employers, government regulators, and users of health care services. Supply chain practices and roles of the participants have been substantially affected by the Affordable Care Act 2010 as focus switches from volume to efficiency (Banker 2014).

This chapter describes trends, issues, and some solutions for logistics management in health care supply chain with concepts drawn from Industrial Engineering (IE), and Operations Research (OR) disciplines applied to specific domains. A health care supply chain model utilizing e-commerce strategy is presented. A hospital laboratory supply chain is used as an example of diverse range of health care supply chains. On-site testing and outsourcing options for supply chain configuration are compared depending on demand patterns. The temporal aspects of health care supply chain are shown to have major impact on configuration decisions and the process perspective is used to describe the supply chain. The example is motivated by interviews conducted with managers at a hospital laboratory. Similar issues from the technology adoption perspective have been analyzed by Jacobsen and Jørgensen (2011).

The rest of the chapter is organized as follows. Section 15.2 makes the case for a Health Care supply chain emphasizing the need for it, strategic drivers, key issues and opportunities that exist for it to be a viable alternative for businesses. An e-health care supply chain model is presented in Sect. 15.3. Section 15.4 explores the hospital laboratory supply chain example. The chapter concludes with suggestions of possible problem areas where the proposed framework can be suitably applied as a future task, described in Sect. 15.5.

15.2 Health Care Supply Chain: The Need, Drivers, Issues, and Opportunities

Supply chain management in the health care industry has many facets and presents ample opportunities for improvement. Health care supply chain networks consist of large number of units albeit majority of them are weakly integrated.

15.2.1 Why Supply Chain Management for Health Care Industry?

There are a number of reasons why the health care industry needs to look at how they manage their supply chain. The main ones—cost and risk. According to Bradley (2000), "how well or badly the health care supply chain is managed is a major factor in health care costs". During the mid-1990s, the Efficient Healthcare Consumer Response (EHCR) (EHCR 2000) performed its own major supply chain study. They found out that the health care supply chain inefficiencies contributed $11 billion (or 48 %) out of the total annual costs of $23 billion. Their report described that the health care supply chain was centered around distributors, resulting in little contact between manufacturers and hospital materials managers. Contract negotiations tended to be adversarial. Providers achieved lower costs, but these costs were not driven out of the system, just pushed lower in the supply chain. Darling and Wise (2010) indicate that improvement of supply chain practices can lead to 2–12 % of savings in hospital operational costs.

15.2.2   Health Care Supply Chain Drivers

The factors that are driving the call for efficiency in health care supply chain are based on common business sense realizing that considering the size of the industry, even small-scale efficiencies can have potentially large dollar impact. Some of the key drivers are described below:

- *Fragmented supplier base.* With 26,000 medical suppliers, managing vendor relationship, costs significant time and money for the buyer. The goal should be to consolidate purchases so as to buy majority of products from one source.
- *Reduced government subsidies* have created the necessity to control costs. In addition, the Health Insurance Portability and Accountability Act of 1996 (HIPAA) and the Affordable Care Act of 2010 (ACA) regulations have created the urgency for providers to address security and electronic transactions issues, resulting in additional cost of doing business.
- *Supply chain inefficiencies.* As mentioned earlier, according to the Efficient Health Care Consumer Response (EHCR) study, approximately $23 billion is spent on the US health care supply chain annually. Streamlining the ordering process and reducing number of supplier relationships can eliminate approximately $11 billion in costs.
- *Managing core competencies.* An efficient supply chain frees up time for health care professionals to focus on their core competency of delivering quality patient care.
- *Internet based purchasing.* This enables supplier consolidation, reduced ordering costs, and a common purchasing platform for hospital networks.
- *Common data standards.* Adopt and promote uniform industry data standards for supply chain transactions over the Internet.
- *Standardization of product purchases in the supply chain.* Standardization of hospital supplies for their impact on (a) purchase volume, (b) ordering and tracking, (c) storage space, (d) resource allocation, and (e) economies of scale through group purchasing power (Vermond 2000).

15.2.3   Integrated Supply Chain Process: Key Issues and Opportunities

There is now a greater awareness in the health care industry that there are significant payoffs through efficient management of the health care supply chain, whose processes incur avoidable costs in following areas:

- Transportation from a production plant to a regional distribution center.
- Distribution center operations.
- Outbound freight.
- Wholesale distributor's receiving and warehousing operations.
- Wholesaler distributor's mark-up for information processing and customer service.
- Transportation to the care provider.
- Inventory.

Integrated supply chain processes would transform this disjoint string of activities into streamlined, cost effective processes characterized by substantial standardization, integration, and optimal service placement (Brennan 1998). In order to successfully integrate the supply chain processes, five supply chain management areas need to be met or exceeded, as per the results of study published by Pricewaterhouse Coopers (cf. Chandra and Kachhal 2004), and summarized in Table 15.1.

For the Health Care Supply Chain, these supply chain management areas are elaborated below.

Demand Management. Managing consumption of clinical resources is key to controlling demand and reducing the number of supplies that move through the supply chain process. Three practices need to be implemented in this regard:

Table 15.1 Supply chain management applications and potential savings for health care in the USA, 2000 (as a % of procurement costs)

Supply chain management area	Potential benefits	Percent of procurement cost
Demand management	Minimized duplication, planning system, demand-driven ordering (clinical guidelines etc.)	2–4 %
Order management	Consolidated purchasing, paperless order management (EDI, Internet)	2.5–4 %
Supplier management	Supplier consolidation, optimal direct-from-manufacturer implementation, compliance with GPO agreements	0.5–2 %
Logistics management	Consolidated service center, integrated transport network, capacity utilization	0.5–2 %
Inventory management	Automated point-of-service distribution, replenishment, non-stock items, reduction in SKUs	0.5–1.5 %
	Overall cost-savings from supply chain management 6–13.5 %	

(1) demand needs to be forecast and a plan implemented to facilitate fulfillment of supplies on a periodic basis, (2) standardization of supplies so as to deliver them as a single unit of inventory, and (3) development of clinical guidelines to define supply requirements for key patient groupings.

Order Management. Initiating effective order management practices:

* Establishing standard order management processes.
* E-procurement through Web or electronic data interchange.
* Implementation of electronic product numbering and tracking process.

Supplier Management. Some of the key ingredients of an effective supplier management process are as follows:

* Reducing the number of suppliers that provide product to the health care system.
* Establishing and participating in group purchasing contracts to take advantage of discounts and rebates.

Logistics Management. Integrated logistics management that exploits efficiencies offered by consolidation of shipments, utilization of service centers and transportation network, and cross-docking in transportation of goods.

Inventory Management. Reducing the storage space, minimizing stock keeping units and their stocking levels, and maximizing inventory turnover rates can achieve integrated management savings. One of the key enabler of this policy is reducing variability among common products through standardization initiatives.

15.3   e-Health Care Supply Chain: Business Trends, Initiatives, and Model

e-health care can be described as the transition of health care business and patient-related processes and transactions into the Internet-delivered electronic information superhighway. The concept of e-health as it evolves, refers to the use of Web-enabled systems and processes to accomplish some combination of following objectives: cut costs or increase revenues, streamline operations, improve patient or member satisfaction, and contribute to the enhancement of medical care (Bose 2003).

According to a study published by Forrester Research, the Internet health care industry in the USA will become a $370 billion business by 2004. Firms will organize around a health care e-business network that will serve consumers, providers, distribution chains and payers (Dembeck 2000). According to this study, "Eight percent of retail health sales will move online." Health e-tailers such as Rx. com and Vitamins.com will experience retail growth to $22 billion in 2004. The real growth in e-health care, however, will be in the business-to-business segment, which the study predicts will soar to $348 billion in 2004. As online business trade gains momentum, 17 % health care business transactions will move online by 2004.

Forrester's analysis of the consumer e-commerce health care market projects healthy growth in online trade for—non-medication health and beauty aids, over-the-counter non-prescription drugs, natural health cures, and prescription drugs.

According to this study, both large institutions and small medical practices will turn to Net players, such as Embion.com, Medicalbuyer.com and Medibuy.com to simplify procurement of medical supplies, thereby driving Internet efficiencies into the distribution chain. As a result, the study predicts that cost-conscious hospitals will move 24 % of their purchasing online by 2004. Meanwhile, as more doctors get connected to health care networks, 12 % of private practices will conduct their procurement online by 2004.

e-health care initiatives. In the health care industry, web-enabled applications under development include products for: claims handling; physician practice management systems; online prescriptions; and electronic clinical and financial data interchange for hospitals, physicians, pharmacies, managed care organizations and commercial and hospital laboratories. Other applications include: patient-centered systems; solutions for chronically ill patients; finance and accounting programs for hospitals and other health agencies; medical supply purchasing; health and medical web portals; managed care organization provider directories; health promotion and disease prevention; provider credentialing; risk management; case management; and practice management.

The move to Internet-based programs and services should result in savings for employers, insurers, managed care organizations and government-sponsored programs because of the significant cost-saving opportunities, such as better price comparisons, lower inventory costs, and more efficient health system-wide communications, patient information management and billing and claims handling (Nugent 2000).

e-health care supply chain model. The emergence of digital business value chains in the health care industry will lead to a trend for its supply chain management. The technology that is already available to integrate Web front-end interactions with back-office systems include, packaged Web modules, "middleware" and tools to build customized transaction systems, Web-based EDI, Web-based electronic marketplaces, and many electronic catalogs. This type of network connection allows the consumer to go directly to the system of choice to design, configure, and arrange for shipment or availability of the final product or service of choice. This ability is the result of an Extranet, linking partners in supply chain to necessary information on-line. An Intranet is used to link technologies, business processes, and organizational constituencies into a network that can display its offerings to consumers over the Internet.

Figure 15.1 depicts a representative model of US e-health care supply chain proposed in Kumar and Chandra (2001), showing linkages and flow (material and information) between various business entities. The principal participants in the US health care supply chain include: manufacturers (drugs, medical equipment, and hospital medical supplies), distributors, medical service providers, medical groups, insurance companies, government agencies (such as, Health and Human Services), employers, government regulators, and users of health care services.

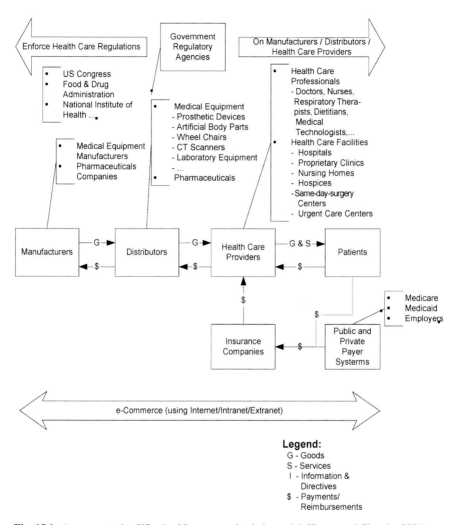

Fig. 15.1 A representative US e-health care supply chain model (Kumar and Chandra 2001)

15.4 Health Care Supply Chain Example

To illustrate some of the complexities of supply chain configuration in the health care industry, an example from the hospital laboratory testing supply chain is investigated. Laboratory tests play a key role in diagnostics and time is one of the most crucial factors in these investigations. Timely diagnostics is crucial to prescribe appropriate treatments. Additionally, laboratory materials as well as test samples are perishable products. Trust and transparency is a must and certification and quality management is required for providing laboratory services. Supply chain configuration depends upon the type of tests to be performed.

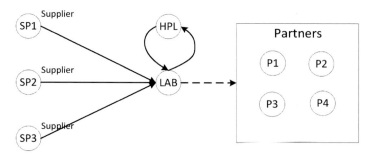

Fig. 15.2 Hospital laboratory supply chain

15.4.1   Supply Chain Description

The hospital laboratory supply chain is shown in Fig. 15.2. It consists of the hospital, laboratory, suppliers of testing materials and partners. The hospital consists of multiple clinical units. The laboratory and the hospital are co-located though travel distances are significant, given that testing samples and materials are often handled on per-order basis. Relative to the laboratory, the hospital is both supplier and customer. It provides testing materials and consumes testing results. The partners are specialized labs performing specific tests and possessing unique skill-set and certifications. The hospital may collaborate with independent laboratories though it puts more trust in on-site testing. This example focuses on the hospital laboratory and interactions with related labs are beyond the scope of the example presented.

   Given importance of temporal factors, the process perspective of the conceptual modeling is of particular interest. The process model (Fig. 15.3) shows interactions among the supply chain units involved. Suppliers are selected via a tender. The laboratory itself grows microorganisms necessary in investigations and the micro-organisms have finite, relatively short lifespan. Some of the microorganisms are nurtured continuously while others are created on demand. The laboratory samples and testing results are exchanged between the hospital and the lab using pneumatic tubes. The laboratory outsources test because it does not possess the required qualifications or lacks materials and capacity. That is especially true for epidemi-ological diseases like salmonellosis (Horby et al. 2003). The hospital lab does not find it feasible to maintain necessary materials and to acquire certifications for such rare occurrences.

15.4.2   Experimental Evaluation

The supply chain configuration objective is to determine the right kind of supply chain structure depending on the diagnostics type where configuration alternatives are: (1) on-site investigations requiring setting-up new tests and undergoing

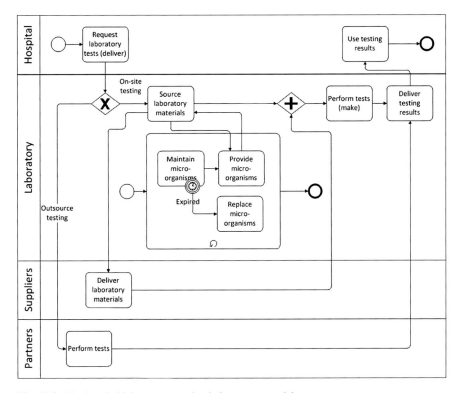

Fig. 15.3 The hospital laboratory supply chain process model

certification; and (2) outsourcing to partners. Outsourced tests are typically more expensive and involve extra transportation costs while on-site tests might require additional certification, training and equipment.

Demand for different types of diagnostics and testing varies and some of the typical demand patterns are shown in Fig. 15.4. These patterns include tests having stable demand and test having irregular demand where demand spikes are associated with outbreaks of particular diseases. The demand is characterized by four parameters: (1) base demand; (2) outbreak occurrence frequency or interval between outbreaks; (3) intensity of outbreaks; and (4) duration of outbreaks. There are tests having near-zero demand in absence of outbreaks (Fig. 15.4c).

It is expected that a choice between on-site testing and outsourcing depends upon the demand pattern. Therefore, a simulation study comparing two supply chain configuration is performed. The testing cost over the fixed planning horizon is used as an evaluation criterion. In the case of on-site testing, the testing cost C_s consists of the fixed setup cost for certification, training and equipment, the variable cost per test performed and the cost for nurturing microorganisms and their replacement after expiry (regardless whether used or not). In the case of outsourced testing, the testing cost C_o consists of transportation cost per small batch of testing

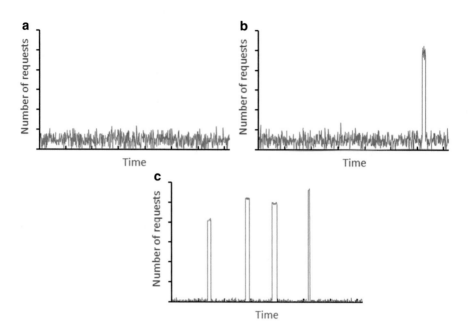

Fig. 15.4 Demand patterns: (**a**) stable demand; (**b**) rare outbreaks; and (**c**) regular outbreaks with low basis demand

samples and the variable cost for every test performed. The variable cost per test is 25 % more expensive at the partner site than on-site. The configurations are compared using the costs ratio $R = C_s/C_o$.

Demand for the tests is simulated and three experimental factors are considered. Base demand, interval and duration are varied to have either low or high values (actual values used in the study are illustrative rather than empirical observations) and all possible combinations of these factors are considered. Hundred simulation replications are executed for every experimental treatment. The average number of outbreaks varies from 0.77 to 1.78 outbreaks over the 2 year period depending on the experimental treatment.

Results of experimental evaluation are reported in Fig. 15.5. Values of $R < 1$ indicate that the on-site testing configuration is preferred. The most significant factor is the base demand. The results show that on-site testing is beneficial only if there is sufficient regular demand for tests. The outbreak frequency also significantly influences the configuration choice in favor of the on-site testing while the duration within boundaries considered does not affect the choice. The simulation study yields a decision rule that if the base demand is at the low level then the outsourcing configuration is the preferred configuration and partners should be involved in testing. There is some evidence that the on-site testing configuration could be adopted in the case of regular outbreaks (i.e., the interval factor is at the low level). Indeed, if outsourced testing premium was increased to 50 % then the on-site testing became the best option in the case of regular outbreaks even for low base demand (results are not shown).

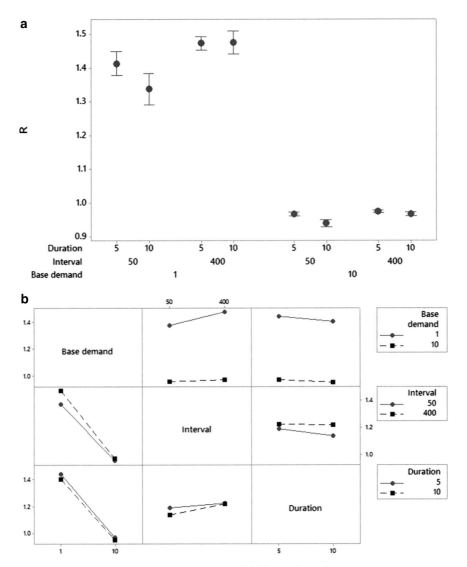

Fig. 15.5 Evaluation results: (**a**) interval plot; and (**b**) interactions plot

15.5 Conclusions

The chapter provides an overview of issues in health care supply chain management including an example of the hospital laboratory supply chain. It has been observed that health care supply chains are enormous, though weakly integrated networks. This provides ample opportunities for improving health care supply chain management and reconfiguring the network structures with much stronger emphasis on collaboration. Given the scale of supply chain networks, the network analysis techniques are likely to find applications in the health care supply chains.

Health care supply chains can be perceived as hybrid supply chains because of the variety of entities involved. They have characteristics of product-oriented as well as service-oriented supply chains. The items processed by health care supply chains are often invaluable; for instance, in the case of heart transplantation when supply chain needs to be configured on-demand within a limited time window. Supply chain units have vastly different roles and are often highly specialized. The networked behavior is more profound than in the traditional industrial supply chains. Many of the health care supply chain management activities and communication links are increasingly digitized and the concepts of cloud chains and solution supply chains will be highly relevant. Time factor is of crucial importance since timely diagnostics and treatment might prevent further progression of diseases. It is also the case of the hospital laboratory where intermediate results of long running tests are transferred to the hospital using an integrated information system. Partners willing to provide similar services also should be able to become a part of the integrated system.

References

Banker S (2014) Obamacare transforms the hospital supply chain. Forbes, 18 Feb. http://www.forbes.com/sites/stevebanker/2014/02/18/obamacare-transforms-the-hospital-supply-chain/

Bose R (2003) Knowledge management-enabled health care management systems: capabilities, infrastructure, end decision-support. Expert Syst Appl 24:59–71

Bradley P (2000) Taking on the health care supply chain. MSI 50

Brennan CD (1998) Integrating the healthcare supply chain. Healthc Financ Manage 52(1):31

Chandra C, Kachhal SK (2004) Managing health care supply chain: trends, issues, and solutions from a logistics perspective. Proceedings of the sixteenth annual society of health systems management engineering forum, Orlando, FL, 20–21 Feb, 2004

Darling M, Wise S (2010) Not your father's supply chain. Following best practices to manage inventory can help you save big. Mater Manag Health Care 19(4):30–33

Dembeck C (2000) Online healthcare expected to reach $370B by 2004. E-Commerce Times 4:1–2

EHCR (2000) Efficient Healthcare Consumer Response Assessment Study. EHCR

Horby PW, O'Brien SJ, Adak GK, Graham C, Hawker JI, Hunter P, Lane C, Lawson AJ, Mitchell RT, Reacher MH, Threlfall EJ, Ward LR (2003) A national outbreak of multi-resistant Salmonella enterica serovar Typhimurium definitive phage type (DT) 104 associated with consumption of lettuce. Epidemiol Infect 130(2):169–178

Jacobsen P, Jørgensen P (2011) Improving hospital logistics by rethinking technology assessment. 12th International CINet conference: continuous innovation: doing more with less, pp 358–372

Kaplan GS (2012) Waste not: the management imperative for healthcare. J Healthc Manag 57(3):160–166

Kumar S, Chandra C (2001) e-Health care: a novelty or a vehicle for change. Ind Eng 33(3):28–33

McKone-Sweet KE, Hamilton P, Willis SB (2005) The ailing healthcare supply chain: a prescription for change. J Supply Chain Manag 41(1):4–17

Nugent D (2000) E-commerce: an Rx for organizational pain. Health Manag Technol 21 (10):32–37

Vermond K (2000) Supply chain management in health care: the OHA taskforce report

Chapter 16
Future Research Directions in Supply Chain Configuration Problem

16.1 Introduction

As in the case of any open and adaptive system, the structure of supply chain has evolved progressively over time from a sequential supply chain, to a global supply chain, a supply network, and alliance networks, respectively. This evolution has reflected the change in business environment from static to dynamic. In Chaps. 1 and 2, we discuss such supply chain configuration phenomenon and shed light on its sources and causes. It was also observed that various components of supply chain have a significant impact on its structure. So what does the future hold for the supply chain configuration problem? To answer this question, we review the anatomy of a supply chain from the perspective of trends and opportunities and their impact on its structure. Then, we propose an agenda for future research in supply chain configuration, which takes into account the confluence of interdisciplinary research and the increasing use of emerging technologies.

16.2 Trends and Opportunities in Supply Chain Configuration

To prognosticate trends and opportunities in the area of supply chain configuration, one only has to review some of the vital issues driving the development of manufacturing and logistics in the twenty-first century (Lasi et al. 2014; NRC 1998):

- Increasing consumer expectations for customized products has led to demand segmentation and fragmentation.
- Next wave of globalization has changed traditional regional division of labor in international supply chains.

© Springer Science+Business Media New York 2016
C. Chandra, J. Grabis, *Supply Chain Configuration*,
DOI 10.1007/978-1-4939-3557-4_16

- Environmental issues are addressed in a more holistic manner rather than just focusing on individual aspects such as carbon emissions. That has increased importance of supply chain traceability and accountability.
- Resilience to withstand random disturbances in the inter-connected world has come to the forefront of supply chain management implying that efficiency without flexibility is not sufficient.
- Products as well as supply chain processes are increasingly digitized, thus blurring boundaries between the physical and digital products and operations.
- Information technologies, such as cloud computing provides ample computational power for solving complex configuration problems and sharing data. Internet of Things allows for capturing data at their origination and mobile technologies enable easy communication (Tien 2015).

Mass customization enabled production of customized products efficiently. However, nowadays customers often become a part of product development, manufacturing, and delivery activities (Wu et al. 2013). For instance, additive manufacturing allows customers to print spare parts. Demand is also increasingly fragmented because new groups of customers are appearing and products are consumed in increasingly specific contextual situations.

Traditional patterns of supply chains have witnessed various forms of transformation. For instance, South-East Asia has long been a region supplying components and performing contract manufacturing in the electronics industry for supply chains serving customers in the USA and Europe. Nowadays, customers are scattered around the world and Asian companies increasingly develop and market their own products. Companies assume different roles in supply chains and products flow in different directions. At the same time, customers require greater transparency in supply chains which is more difficult to attain in networked supply chains. Governmental regulatory requirements are often superseded by a variety of informal guidelines of ethical behavior (Yusuf et al. 2014).

Strong emphasis on cost reduction and lean practices has been challenged by a string of exceptional events ranging from natural disaster to man-made calamities. Supply chain evaluation according to multiple-criteria and risk management emerges as one of the critical supply chain management areas (Heckmann et al. 2015).

Development of information technologies affects supply chains in different ways. The physical products are packaged together with digital products and services, and costs for opening warehouses are replaced by costs associated with establishing digital distribution avenues involving channel development, marketing; including social marketing and establishing integrated business models. Contractual issues are becoming particularly important for the combined physical–digital products since revenues and liabilities are often attained indirectly and over prolonged periods of time. On the other hand, information technologies play a major role in increasing supply chain flexibility by reducing communication cost, transforming delivery channels, and providing instantaneous access to data about supply chain process, products, and customers.

16.3 Future Research Agenda

The aforementioned trends motivate future supply chain configuration research agenda. The increasing customer expectations and the digital–physical blur can be addressed by focusing on solution supply chain rather than product-oriented supply chains. Networked and efficient logistics clusters capture opportunities provided by globalization and clusters act as multipurpose units in global supply chains. New business models are needed in the solution supply chains and environmentally and socially conscious supply chains. Big data as the basis for big modeling contributes to development of resilient supply chains.

The increasing role of information technologies and digital products in supply chains has the potential of creating unique capabilities for improving supply chain management. Therefore, it is befitting to recognize that it will have a prominent role in defining the agenda for future research in supply chain configuration. As described in previous chapters, information sharing and information integration are two of the key problems in supply chain management. As the size of the supply chain network grows, there is an exponential increase in the amount of data—and, therefore, information and eventually knowledge—that needs to be acquired, stored, managed, processed, and serviced for various decision-making needs, while managing the supply chain. This problem needs efficient solution both from an operational perspective (forecasting and inventory management), as well as development of efficient information processing methodologies and techniques. *Cyberinfrastructure* offers that venue for supply chain configuration research.

More specifically, the key dimensions to be considered for future research in supply chain configuration area are as follows:

Design of Problems. As we have elaborated in previous chapters, supply chain configuration can have potentially a complex web of problems, which may have to be dealt at different decision-making levels. These problems need to be coordinated to design efficient problem-solving solutions. Therefore, the design of new supply chain solutions must account for appropriate relationships among these problems. For instance, product design and environmental issues are well embedded in the current supply chain configuration models while collaborative business models underlying the supply chain partnership is a future research area.

Design of Solutions. The development of solutions for modification to supply chain information systems, and the integration of advanced decision-making components by adopting modern software engineering techniques, offers opportunities for better supply chain management. Big data increasingly finds application in managerial decision-making. Big modeling (i.e., large-scale integration of decision-making models) is the next step in design of solutions (Tolk 2014). It would enable evaluation and comparison of large number of supply chain evolution scenarios from multiple perspectives.

A System-of-Systems Approach is needed to design complex supply chain networks. This is especially true as supply chains assume global proportions, whereby the number of entities and their relationships multiply disproportionately. It also recognizes the concept-to-fruition notion of product and service delivery, which in the case of a supply chain may sometime span several heterogeneous and independent systems, and must be integrated together to be effective in delivering the product. The systems approach could emerge as one of key techniques for dealing with globalization and resilience related challenges.

Standardization and Interconnectedness. Adopting standards facilitates meta-modeling and implementation of complex networks, thereby saving on development time for a system. The Supply Chain Operations Reference (SCOR) model is an example of successfully applying industry process standards to conceptual modeling of supply chain networks (Stewart 1997). Similarly, web technologies have contributed to achieving Interconnectedness. Data driven and decision-enabled supply chain processes (Deokar and El-Gayar 2011) require further integration especially concerning exchange and joint utilization of decision-making data.

Design-Time and Run-Time Reconfigurability of Supply Chain. Design-time reconfiguralibility is required to deal with customer demand fragmentation, short life-cycle products, small series production and resilience issues. A new or revised existing customer requirement for a product specification may spawn modifications and/or enhancements, with potential impact on the physical and logical systems enabling the realization of the product. The ramifications of such changes should be considered during the conceptualization phase in product design. Changes in product specification typically affect the essential ingredients for competitiveness—namely, minimal cost, lead time, and optimal product variety. The complexity of changes in such a system is magnified when the design, manufacture, and logistics of the product is accomplished in a distributed environment. The contextual information plays an important role under these circumstances. Context aware supply chains capable of changing their behavior in run-time in response to changing context are one of the future solutions to attain reconfigurability.

Solution supply chains. The cloud chain concept introduced in Chap. 12 demonstrates integration between physical and digital units in supply chains. Similarly, supply chains increasingly serve their customers by providing solutions (Cavalieri and Pezzotta 2012) combining physical and digital products as well as services. This kind of supply chains can be referred to as solution supply chains and the main focus of their configuration is development of the right product–service mix and establishing mutually beneficial relationships in the network of solution providers. Solution supply chains are likely to be supported by cyberphysical supply chain processes similarly as in the case of Industry 4.0 (Pisching et al. 2015).

The agenda for future research in supply chain configuration must recognize the urgent need for providing solutions that particularly satisfy public policy applications. A generic example of this is the design of supply chain configuration for the *unknown*, which may be any of the following applications:

- Geopolitical issues and cross-country relationships.
- Community issues affecting local (i.e., region, city), state, country, and global levels.
- Mega-disasters, including man-made and natural disasters.

- Cybersecurity.
- Disease prevention and control.
- Management of environmental issues.
- Public finance issues.
- Energy consumption and preservation issues.

These issues are tightly related with reverse logistics, green supply chains, long-term sustainability, and supply chain transparency. Concerning sustainability, a major concern is balancing the typical 2–10 years planning horizon with environmental and social sustainability targets having substantially longer planning horizons. Supply chain configuration tools could be a decision-making tool not only for commercial enterprises, but for governmental institutions to investigate potential impact of particular regulatory requirements.

Other future areas of research are improving computational capabilities of supply chain configuration models and techniques. In this regard, an important area is elaboration of methods for integration of models in decision modeling, and models in information systems design, to offer integrative decision-modeling capabilities. These could certainly be tied to research in cyberinfrastructure to recognize an inter-operable environment offered by the Internet.

References

Cavalieri S, Pezzotta G (2012) Product-service systems engineering: state of the art and research challenges. Comput Ind 63(4):278–288

Deokar AV, El-Gayar OF (2011) Decision-enabled dynamic process management for networked enterprises. Inform Syst Front 13(5):655–668

Heckmann I, Comes T, Nickel S (2015) A critical review on supply chain risk – definition, measure and modeling. Omega 52:119–132

Lasi H, Fettke P, Kemper H, Feld T, Hoffmann M (2014) Industry 4.0. Bus Inform Syst Eng 6 (4):239–242

National Research Council (1998) Visionary manufacturing challenges for 2020. Committee on visionary manufacturing challenges, National Research Council. National Academy Press, Washington, DC

Pisching MA, Junqueira F, Filho DJS, Miyagi PE (2015) Service composition in the cloud-based manufacturing focused on the industry 4.0. Technological innovation for cloud-based engineering systems. IFIP Adv Inform Commun Technol 450:65–72

Tien JM (2015) Internet of connected ServGoods: considerations, consequences and concerns. J Syst Sci Syst Eng 24(2):130–167

Tolk A (2014) M&S as a discipline – foundations, philosophy, and future. In: Tolk A, Diallo SY, Ryzhov IO, Yilmaz L, Buckley S, Miller JA (eds) Proceedings of the 2014 winter simulation conference, pp 3995–3996

Stewart G (1997) Supply-chain operations reference model (SCOR): the first cross-industry framework for integrated supply-chain management. Logist Inform Manag 10:62–66

Wu D, Greer MJ, Rosen DW, Schaefer D (2013) Cloud manufacturing: strategic vision and state-of-the-art. J Manuf Syst 32(4):564–579

Yusuf Y, Hawkins A, Musa A, El-Berishy N, Schulze M, Abubakar T (2014) Ethical supply chains: analysis, practices and performance measures. Int J Logist Syst Manag 17(4):472–497

Index